P9-CRK-156

KNIFE RIVER

TTLE BIGHORN
TTLEFIELD

HORN
CINE WHEEL

IROQUOIAN
ARCHAELOGY
IN SOUTHERN
ONTARIO

SERPENT
MOUND

HOPEWELL

CAHOKIA

BLACKWATER DRAW

SPIRO

MOUNDVILLE

POVERTY POINT

Little Bighorn Battlefield

Knife River

Iroquoian Archaeology in Southern Ontario

Moundville

Spiro

Cahokia

Pueblo Grande

Pueblo Bonito

Mesa Verde

Serpent Mound

Poverty Point

Hopewell

Ozette

Big Horn Medicine Wheel

Head-Smashed-In Buffalo Jump

Cape Krusenstern

Hidden Cave

| 3,000 B.C. | 2,000 B.C. | 1,000 B.C. | A.D. I | 1,000 A.D. | 2,000 A.D. |

Exploring Native North America

PLACES IN TIME

Series editors Brian M. Fagan and Chris Scarre

PLACES IN TIME

Exploring Native North America

David Hurst Thomas

OXFORD
UNIVERSITY PRESS
2000

OXFORD
UNIVERSITY PRESS

Oxford New York

Athens Auckland Bangkok Bogotá Buenos Aires Calcutta
Cape Town Chennai Dar es Salaam Delhi Florence Hong Kong Istanbul
Karachi Kuala Lumpur Madrid Melbourne Mexico City Mumbai
Nairobi Paris São Paulo Singapore Taipei Tokyo Toronto Warsaw

and associated companies in
Berlin Ibadan

Published by Oxford University Press, Inc.
198 Madison Avenue, New York, New York 10016-4314

Library of Congress Cataloging-in-Publication Data
Thomas, David Hurst.
        Exploring Native North America / David Hurst Thomas.
            p. cm.—(Places in time)
        Includes bibliographical references and index.
        ISBN 0-19-510887-6
            1. Indians of North America—Antiquities. 2. North America—Antiquities.
    I. Title. II. Series.
    E77.9.T5 2000
    970.01—dc21   99-056046

9 8 7 6 5 4 3 2 1

Printed in the United States
Designed by Diane Gleba Hall

Contents

Preface

Indian people have lived in North America for a thousand generations, perhaps longer—perhaps much longer.

This book provides an up-to-date introduction to the archaeology of Native North America, written for a fairly broad readership including high school and college students, heritage tourists, and avocational archaeologists. My job as a professional archaeologist is to boil things down to basics, to distill for you the essence of what I see as consequential in American archaeology. I will try to provide you an authoritative, current introduction to some key archaeological sites in North America. These are all places that you personally can visit, should you wish.

This Is Not a Book about Prehistoric Indians

When this project began, I was asked to write a book in the *Places in Time* series, sponsored by Oxford University Press. The subject matter was projected as the archaeology of "Prehistoric North America."

But I was uncomfortable with this concept, for a couple of reasons. Having worked with Native Americans over the years, it was immediately clear to me that the entire notion of "prehistoric" is problematic.

Most Indian people that I know don't like having their ancestors classified as "prehistoric," as somehow "before history"—or worse yet, "without history" at all. I have heard Indians comment that, to them, the term prehistoric conjures up comparisons with dinosaurs, flightless birds, and other extinct critters. Understandably, they do not appreciate these comparisons. Indian people have increasingly taken the position that they have always had their own history, but their own views have traditionally been ignored or denigrated by white American scholars. This is why the idea of a book about the archaeology of prehistoric North America did not work, at least not for me.

The editorial staff at Oxford University Press was sympathetic to this problem, and they provided me considerable leeway in developing a different approach. You have in your hands a book defined simply as being about Native North American archaeology. I discuss eighteen key archaeological sites, carefully chosen to reflect the long-term history of American Indian people, from the 11,000-year-old Clovis occupation at Blackwater Draw (New Mexico) through the Battle of the Little Bighorn (25 June 1876).

Indian people have been in America for thousands and thousands of years. They are still here in great numbers. They have always had a history and they always will. The term prehistoric does not appear here because the concept is irrelevant to the story I wish to tell.

Site Selection

North America is one immense outdoor museum, telling a story that covers 9 million square miles and 25,000 years. America is a land of hands-on historians, people who want to get out and see the real thing. And this is what museums have traditionally always offered—a chance to experience the real thing.

I take a pretty broad view of what a museum is. To me, an "archaeology museum" is anything that publicly presents something important from the past. America contains thousands of such museums—from the largest urban natural history museum to major archaeological sites like Cahokia and Chaco Canyon. But we cannot forget those small roadside attractions signaled by a fading roadside sign saying simply, "Point of Historical Interest, ½ mile ahead." These are archaeology museums as well.

To be included here, an archaeological site must, above all else, have major historical, cultural, or methodological significance. That site must also be readily available to the traveling public: all such places must encourage visitation, provide interpretation, and ensure adequate protection for both the visitor and the archaeological record remaining at the site. The idea is to protect the sites from the public and the public from the sites.

Beyond this, we have imposed a chronological, geographic, and conceptual balance in picking sites for inclusion. The contents are also deliberately slanted toward places where archaeology has made a particularly significant contribution to our current understanding of what happened.

This means that sites are not included for purely historical reasons—simply because "something important once happened there." The term "important" is culturally loaded, and because I am deliberately embracing a multicultural perspective, some places important to one group or another will inevitably be left out.

Some are well-known places, major tourist attractions such as Mesa Verde, Pueblo Bonito, Cahokia, and the Little Bighorn Battlefield. Other sites are regional attractions, places such as the Knife River villages, Spiro, Pueblo Grande, Serpent Mound, and Poverty Point. A few places—such as Blackwater Draw, Hidden Cave, and Cape Krusenstern—are well-kept secrets, but archaeologically significant just the same.

Sacred Sites and the American Public: Not an Easy Call

When talking about Cahokia or Mesa Verde, tourism is a given. Archaeological sites like these are important elements in our cumulative national identity, and people will always be motivated to experience the power of such places for themselves. Closing the Little Bighorn Battlefield to tourists would be sacrilegious to Anglo-Americans, for whom such historic touchstones provide information about our national, heroic roots. The Native American community also wants people to visit the Little Bighorn Battlefield, but they want to present a rather different message there. Archaeologists have a role to play for both constituencies.

You will note that I have included Wyoming's Big Horn Medicine Wheel as one of our destination sites, raising in the process a rather different set of concerns. Many archaeologists, myself included, believe that the Big Horn Medicine Wheel is one of America's most intriguing ancient sites. With its spectacular setting and puzzling past, the Medicine Wheel is a natural for the heritage tourist. But to many Indian people, the Medicine Wheel is also a sacred site, a holy place where important ceremonies are performed to this day. Sacred lands are considered vital to individual and tribal harmony.

Some of North America's most important sacred sites—including the Big Horn Medicine Wheel—are being overrun each year by thousands of non-Indians: curious tourists, Indian buffs, New Agers seeking their own spiritual experiences, and even the occasional team of archaeologists. There is great concern that the plants, paths, shrines, rocks, and other aspects of such sacred sites are being destroyed by the curious and the insensitive.

Should tourists be encouraged to visit the Big Horn Medicine Wheel? How do archaeologists balance the dual concerns of bringing American archaeology to the interested public while respecting the wishes of the descendant populations still involved with many of those sites?

Let me briefly address these important issues by using the Big Horn Medicine Wheel as a concrete example. In their excellent book *America's Ancient Treasures* (4th edition, Albuquerque: University of New Mexico Press, 1993), Franklin Folsom and Mary Elting Folsom describe thousands of archaeological museums and sites suitable for tourist visitation. They provide a first-rate, encyclopedic overview, one designed "to open doors to those who are curious and who want to dig metaphorically into the past" (251). I like this book very much, and for decades have kept the latest edition in my glove compartment.

The Folsoms considered the sacred site issue carefully, and decided to exclude the Medicine Wheel from their guidebook. Noting that they had discussed the issue with representatives at the American Indian Rights Fund, they declined to discuss the site and refused to provide directions for visitation. Instead, they simply wrote, "[The Big Horn Medicine Wheel] is sacred to Native Americans who worship there. They request that visitors stay away and do not invade their privacy" (251). This position reflects a sensitivity to Native interests. They made a good decision.

This book covers some of the same ground as the Folsoms' guide, albeit from a very different perspective. Both books try to educate the traveler about the ancient Native American past. Both also acknowledge the sometimes conflicting aims of archaeological research and traditional American Indian religious beliefs.

But after struggling with the issue of sacred sites I came to the opposite conclusion. In Chapter 5, I describe the Big Horn Medicine Wheel in some detail. I provide you with several possible explanations of its origin and use, as well as precise directions for those wishing to visit.

Like the Folsoms, I consulted various Native people, eliciting a broad range of opinion. A few expressed the view that sacred sites should never be visited by the non-Indian public. Others saw no particular problem with tourists visiting such sites. After all, they pointed out, places like the Medicine Wheel are already visited by thousands of non-Indians every year; no single book or author can hope to stem this flood of visitation. The key issue, they pointed out, is to educate the public about Native values and to be certain that such sacred sites are protected from looting and despoliation.

I was still undecided about whether or not to include the Medicine Wheel when I discussed the matter with the late William Tallbull, a Northern Cheyenne elder with a deep and long-lasting personal relationship to the Medicine Wheel. Tallbull was an important part of a coalition of tribal, scientific, ecological, and government agencies cooperating to ensure that the Medicine Wheel would be protected, preserved, and respected. He supported the Forest Service's decision to keep the site open to the public but with the last 1.5 miles (2.5 kilometers) of the access road closed to vehicular traffic. Tourists are still permitted to visit the site, but are required to cover the remaining distance on foot. In his view, this solution minimized the negative impacts of tourism while maintaining the religious freedom of the Native people (himself included) who worshipped there.

To Tallbull, education remained a key consideration. He also felt it important to keep the site accessible to anybody who wished to experience the power of place. After all, people have been drawn to this isolated mountaintop for centuries, and Tallbull believed it inappropriate to exclude anybody. Not only did he encourage me to include the Big Horn Medicine Wheel in this book, but he offered to write a sidebar to my own discussion. In his perspective, sacred sites offer an important opportunity for teaching tolerance and respect.

Which approach is "correct"? As we encourage people to learn more about North America's archaeological heritage, what should we do about sacred sites? I honestly don't know.

I have no quibble with the Folsoms' decision to exclude the Medicine Wheel; they did so for the right reasons. But I'm also glad to be able to include it here, juxtaposed with the perspective from a respected tribal elder.

A Comment on Further Reading

At the end of each chapter, I provide a short listing of suggested readings. These sources provide additional information about the site in question, and also some background giving a broader context to the issues being discussed. On occasion, I throw in a technical discussion or two, so that the interested reader can follow the topic in more detail. But in most cases, the suggestions for further reading are written for a general, nonspecialist audience.

Acknowledgments

I sincerely thank my research staff at the American Museum of Natural History, each of whom helped out in numerous ways: Margot Dembo, Lorann S. A. Pendleton, Lisa Stock, and Niurka Tyler.

I also greatly appreciate the efforts of the following friends and colleagues, each of whom has read and commented upon one (or more) draft chapters included in this book: Stanley Ahler (PaleoCultural Research Group, Flagstaff, Arizona), Kenneth Ames (Portland State University), Douglas Anderson (Haffenreffer Museum of Anthropology, Brown University), Gerard Baker (Mandan-Hidatsa; National Park Service, Chickasaw National Recreation Area), Todd Bostwick (City Archaeologist, City of Phoenix), Cory Breternitz (Soil Systems, Inc., Phoenix, Arizona), Jack Brink (Archaeological Survey of Alberta), David Brose (Schiele Museum of Natural History), Ernest Burch (Camp Hill, Pennsylvania), Ken Carson (Facility Manager, Head-Smashed-In), Vine Deloria, Jr. (Sioux; University of Colorado), Thomas Emerson (Illinois Transportation Archaeological Research Program, University of Illinois,

xi
—

Urbana-Champaign), Bob Gal (National Park Service, Kotzebue, Alaska), Jon Gibson (University of Southwestern Louisiana), James Judge (Fort Lewis College), Vernon James Knight (University of Alabama), Bradley T. Lepper (Ohio Historical Society), Ricky Lightfoot (Crow Canyon Archeological Center, Cortez, Colorado), William Lipe (Washington State University and Crow Canyon Archaeological Center), John Milner (Pennsylvania State University), Timothy Pauketat (State University of New York, Buffalo), Robert Pearce (London Museum of Archaeology), Brian O. K. Reeves (Lifeways of Canada, Ltd., Calgary, Alberta), Stephen R. Samuels (Bureau of Land Management, Coos Bay, Oregon), Joe Saunders (Northeast Louisiana University), Douglas Scott (National Park Service, Lincoln, Nebraska), Dean Snow (Pennsylvania State University), Vin Steponaitis (University of North Carolina), Thomas Thiessen (National Park Service, Lincoln, Nebraska), David Wilcox (Museum of Northern Arizona), and Thomas Windes (National Park Service, Santa Fe, New Mexico).

I am particularly grateful to the late William Tallbull (Cheyenne) for permitting me to include his comments in my chapter on the Big Horn Medicine Wheel.

Introduction

This book is about the earliest Americans: Who were they? Where did they come from? And what became of them?

These questions have puzzled Euro-Americans since that first fateful Columbian encounter. At the time, most Europeans understood the Bible not only as sacred scripture, but also as an infallible historical record of humankind. Enormous confusion and speculation arose to account for the newly discovered Native Americans.

The mystery deepened with each new discovery of American ruins. How to account for the thousands of prehistoric earthen mounds that dot North America east of the Mississippi? The Spanish explorer Hernando de Soto correctly surmised that many of the mounds served as foundations for priestly temples, but his astute observation was soon lost in a flood of fanciful interpretation. There was the "Lost Tribe of Israel" scenario, which pointed to alleged Native American-Semitic similarities. The fabled Island of Atlantis was seriously proposed by some as the ancestral homeland of the Native Americans. Even voyaging Egyptians and Vikings were cited as hypothetical proto-Americans.

Seeking the First American

We can trace the roots of modern scientific thinking on Native American origins to a remarkably prescient Jesuit missionary, José de Acosta, who first suggested that American Indians came from a Siberian homeland. Writing in 1589, Acosta speculated that small groups of hunters, driven from their Asiatic homeland by starvation or warfare, followed now-extinct beasts across Asia into America millennia before the Spaniards arrived in the Caribbean. To support his theory, he noted that such a journey would require "only short stretches of navigation"—an extraordinary premise given that Europeans would not "discover" the Bering Strait for another 136 years.

Throughout the nineteenth century, a number of well-publicized archaeological finds suggested that Ice Age people may have arrived in America sometime during Pleistocene times (a geological epoch that began two million years ago, during which giant mammals roamed across a landscape characterized by massive ice sheets).

But the evidence for an Ice Age human population in America was not entirely convincing, and two distinct schools of thought emerged. One group accepted the evidence at face value, arguing that during the late Pleistocene (sometime within the last 25,000 years or so), Ice Age people must have hunted the giant game animals that once populated America. With few exceptions, these "early man advocates" were amateur relic hunters or fossil collectors—impassioned and committed, yet poorly versed in the scientific methods and theories of the day.

These crusaders for the Ice Age were countered at every turn by the big guns of nineteenth-century American archaeology, establishment men brandishing advanced degrees, mostly curators and professors at places like the Smithsonian Institution, the Universities of Chicago and California, and the American Museum of Natural History. These learneds contended that American Indians were relative newcomers in America—probably arriving no earlier than the Moundbuilders, Pueblo Indians, and Classic Maya—all of whom left easy-to-recognize archaeological traces dating from the last few thousand years.

At the dawn of the twentieth century, it was obvious to all that better sites must be found: the bones of late Ice Age animals had to be scientifically investigated by trained observers. If any solid evidence could be found—that is, undisturbed associations of artifacts with extinct animal bones, in sediments of Pleistocene age—these associations should be presented to a range of scientific witnesses for verification.

This is exactly what happened at Wild Horse Gulch, about 8 miles (13 kilometers) west of Folsom, New Mexico. In 1926, paleontologists from the Colorado Museum of Natural History in Denver were excavating bison bones buried beneath a layer of clays and gravels several feet deep. Because this kind of bison was known to have been extinct for thousands of years, they were shocked to find two pieces of chipped flint—clearly artifacts of human manufacture—lying loose in their spoil dirt. Once alerted, the paleontologists soon found another artifact, this one still in its original position, imbedded in the clay surrounding a bison rib.

Excited by the association of human artifacts with extinct Pleistocene animals, J. D. Figgins, the museum's director, could hardly wait to share his news with colleagues: People had been in the New World far longer than reputable scientists believed possible. But Figgins ran into a solid wall of doubt and disbelief. Maybe the strata had been jumbled in the geological past. Maybe the excavators inadvertently mixed ancient strata themselves. Maybe the flaked stones were much later than the bones, having fallen into the excavation from above.

Still believing in his Folsom finds, Figgins vowed to find better evidence the following year. The 1927 expedition to Folsom was jointly sponsored by the Colorado Museum of Natural History and New York's American Museum of Natural History. Before long, the excavators came across additional ancient artifacts—wonderfully crafted spear points still lodged against the ribs of extinct bison. Having learned from previous disappointments, Figgins and the others left everything in the ground, exactly as they found it.

Telegrams went out, announcing the discovery to the leaders of American archaeology. In a matter of days, scientists began flocking to Folsom to see firsthand the revolutionary finds. This time, the scientific evidence was conclusive.

Virtually overnight, the controversy over Ice Age humans in America evaporated. As it turned out, those sparsely trained, obstreperous amateurs arguing for a long chronology extending back into the Pleistocene were right after all. The elite of American archaeology, critical of all claims for a Pleistocene-era peopling of the New World, were forced to change their minds. Only a month after his visit to Folsom, one highly influential archaeologist, Alfred Kidder, could confidently crow to the world: the first American must have arrived "at least fifteen or twenty thousand years ago."

These earliest Americans are today called *Paleoindians*. Dating between about 11,000 and 6,000 B.C., their archaeological remains have been found from the Pacific to Atlantic shores, from Alaska to the southern reaches of South America. Archaeologists recognize three major cultures within the overall Paleoindian tradition:

> Clovis culture (11,200–10,200 B.P.)
> Folsom culture (10,800–10,200 B.P.)
> Late Paleoindian (or Plano) culture (10,000–8000 B.P.)

In this book, we examine archaeological evidence that links these Paleoindian pioneers with modern Native American populations.

The Beginnings of American Archaeology

Over the past three centuries, American archaeology has undergone some major theoretical and methodological transformations. Modern archaeology employs a variety of perspectives on the past, and the case studies selected for this book deliberately reflect that diversity. But before getting to the specifics, it is important that you understand something about how that archaeology has been constructed.

Thomas Jefferson (1743–1826), the third president of the United States, is generally considered to be the father of American archaeology. Fascinated by Indian lore since boyhood, Jefferson puzzled over the question of Native American origins. He believed that solving this puzzle required a dual strategy: to learn as much as feasible about contemporary Indian culture and to examine Ancient Indian remains.

Jefferson believed that eighteenth-century Native Americans were in no way mentally or physically inferior to the white races and rejected all current racist doctrines explaining their origins. He correctly reasoned that Native Americans were wholly capable of constructing the prehistoric monuments of the United States. Further, as a result of his training in classical linguistics, he believed that Native American languages held valuable clues to their origins. Jefferson personally collected linguistic data from more than forty tribes and wrote a long treatise on the subject. Reasoning largely from his linguistic studies, he deduced an Asiatic origin for Native Americans.

Then Jefferson took a critical step. Shovel firmly in hand, he proceeded to excavate a burial mound located on his property. Today, such a step seems obvious, but none of Jefferson's contemporaries had thought of resorting to bones, stones, and dirt to answer intellectual issues. Contemporary eighteenth-century scholars preferred to rummage through libraries and archives rather than dirty their hands with hard facts from the past.

Written in the flowery style of the time, Jefferson's account in *Notes on the State of Virginia* provides quite an acceptable report of his investigation. First he describes the data—location, size, method of excavation, stratigraphy, condition of the bones, and artifacts—and then he presents his conclusions as to why prehistoric peoples buried their dead in mounds. He first noted the absence of traumatic wounds, such as those made by bullets or arrows, and also observed the interment of children, thereby rejecting the common notion that the bones were those of soldiers who had fallen in battle. Similarly, the scattered and disjointed nature of the bones militated against the notion that it was the "common sepulchre of a town," in which case Jefferson would have expected to find skeletons arranged in more orderly fashion.

Jefferson surmised, quite correctly, that the burials had accumulated through repeated use, and he saw no reason to doubt that the mound had been constructed by the ancestors of the Native Americans encountered by the colonists. Today, nearly two hundred years after Jefferson's excavations, archaeologists would modify few of his well-reasoned conclusions.

Thomas Jefferson's primary legacy to archaeology is the fact that he dug at all. By his simple excavation, Jefferson elevated the study of America's past from a speculative, armchair pastime to an inquiry built on empirical fieldwork. As a well-educated colonial gentleman, Jefferson understood the importance of exposing speculation to a barrage of facts. The "facts" in this case lay buried beneath the ground, and that is precisely where he conducted his inquiry.

Through the efforts of Jefferson and many others, investigators came to appreciate the considerable continuities that existed between the unknown prehistoric past and the Native American population of the historic period. As such knowledge progressed, profound differences between European and American archaeology became more apparent. While the Europeans wrestled with their ancient flints—without apparent modern correlates—American scholars came to realize that the living Native Americans were indeed relevant to the interpretation of archaeological remains. In the crass terms of the time, to many Europeans, the Native Americans became "living fossils," accessible relics of times past.

So it was that New World archaeology became inextricably wed to the study of living Native American people. While Old World archaeologists began from a baseline of geological time or classical antiquity, their American counterparts began to develop an anthropological understanding of Native America. The ethnology of American Indian people not only became an important domain of Western scholarship in its own right, but the increased understanding of Native American lifeways quickly helped unravel questions about the peopling of the New World.

Evolving Perspectives on American Archaeology

Throughout its history, North American archaeology has undergone a series of important transformations. American archaeology began as a pastime of the genteel rich, people like Thomas Jefferson. In chapters 8 and 15 of this book, you will encounter the redoubtable Clarence B. Moore (1852–1936), who purchased a riverboat—the *Gopher*—to serve as his personal base of archaeological operations. Like most gentlemen of his day, Moore personally financed his own fieldwork during the late 1890s and 1900s. Not until the early twentieth century could "working-class" scholars hope to penetrate the archaeological establishment. Women are also virtually invisible in the early histories of American archaeology. Women were, in truth, making their own contributions to archaeology—mostly as unpaid field- and lab-workers—but they were excluded from traditional communication networks, so their contributions are more difficult to identify.

Through the years, archaeology matured into a professional scientific discipline. C. B. Moore was among the first generation of full-time professional archaeologists. As practicing specialists, most archaeologists from Moore's time onward were affiliated with major museums and universities; eventually, others joined the private sector, working to protect and understand America's long-term cultural heritage. Such institutional support not only encouraged a sense of professionalism and fostered public funding, but also forced universities and museums to care for the archaeological artifacts recovered. By the beginning of the twentieth century, American archaeologists were no longer collectors of personal treasure: all finds belonged in the public domain, available for exhibit and study.

There has also been a distinct trend toward specialization. In Jefferson's time, little was known about archaeology, and a single scholar could be aware of all the relevant data. But by the late nineteenth century, so much archaeological information had accumulated that no single scholar could hope to know everything relevant to the archaeology of the Americas. Although Moore would become the leading authority on southeastern archaeology, he knew relatively little about the finds being made by his archaeological contemporaries in Peru, Central America, and even the American Southwest. By the mid-twentieth century, archaeologists had been forced to specialize still further, and today it is rare to find archaeologists with extensive experience in more than one or two specialized fields.

But perhaps the greatest change has been the quality of archaeologists' training. Although broadly educated in science, literature, and the arts, Jefferson's archaeology was wholly self-taught and largely a matter of common sense. Moore also was untrained in archaeology, his fieldwork unfolding through personal trial and error. By the early twentieth century, most American archaeologists were professionally trained and, almost without exception, well-versed in anthropology.

James A. Ford (1911–1968), whose work at Poverty Point is highlighted in Chapter 8, came of age during the Great Depression. This was a time when archaeologists became deeply concerned with discovering chronological sequences and establishing more complete historical reconstructions. During the Great Depression, as the Roosevelt administration created jobs to alleviate difficult economic conditions, crews of workmen were assigned labor-intensive tasks, including building roads and bridges, and general heavy construction. One obvious make-work project was archaeology, and literally thousands of the unemployed were set to work excavating major archaeological sites. This program was an important boost to North American archaeology, and data from government-sponsored Depression-era excavations poured in at a record rate.

This unprecedented accumulation of raw data created a crisis of sorts, forcing archaeologists to develop new methods for synthesizing and classifying, ultimately creating new regional sequences of culture chronologies. By midcentury, archaeologists had derived a series of sequential stages to characterize the precontact archaeological record across the North American continent:

Paleoindian: initial occupation of the Americas
Archaic: widespread post-Paleoindian, nonagricultural
    adaptation based on a broad spectrum of resources
Woodland: manufacture of distinctive ceramics, incipient
    development of agriculture, and construction of funerary
    mounds; the term *Formative* is usually applied to the farm-
    ing, pottery-making cultures of the American Southwest
Mississippian: sedentary farmers of the interior riverine region
    of eastern North America (A.D. 1000–1650)
Ethnohistoric: Native people encountered by Old World
    populations

Although imperfect, these stages persist today, and provide a satisfactory framework for dividing up the long-term Native American occupation of North America.

Several chapters in this book illustrate how this predominantly cultural-historical perspective works in modern archaeology. We discuss, for instance, how the established principles of stratigraphy have helped archaeologists work out detailed cultural sequences at Blackwater Draw (New Mexico) and Cape Krusenstern (Alaska). The discussion of the Knife River Indian villages (North Dakota) also highlights how archaeologists employ a "direct historical approach," working from the established ethnographic record to reconstruct past behaviors from the archaeological evidence.

Beginning in the mid-1960s, American archaeology shifted its center of gravity toward a more scientific, systemic approach. So-called *processual archaeology* began with an unambiguous scientific premise—that knowledge is acquired through public, replicable, empirical, objective methods. Archaeologists were to formulate theories for explaining basic cultural differences and similarities; rival theories are judged by the same criteria, based on their power to predict and to admit independent testing. By embracing this explicitly scientific framework, cultural materialists reject humanist and aesthetic theories that attempt to explain culture on nonscientific grounds. In this view, the most powerful and pervasive determinants of human behavior are environmental, technological, and economic—the material conditions of existence. Religion and ideology were considered to be "epiphenomena," cultural add-ons with little long-term explanatory value. Several chapters in this book proceed from these basic processual assumptions.

Viewed this way, scientific archaeology attempted to provide hard objective evidence about the past. Politics of the present were considered to have nothing to do with the ancient past, and as scientists, archaeologists avoided passing moral judgments on people of the present or the past.

7

Within the past decade or so, American archaeology has been subjected to withering criticism from within. The so-called *postprocessual* critique puts down attempts by "authority" to speak for "the other"—colonized peoples, indigenous groups and minorities, religious groups, women, and the working class—with a unified voice. Each group has a right to speak for itself, in its own voice, and have that voice accepted as authentic and legitimate. A postmodern spirit of pluralism has penetrated the way in which American archaeology is practiced today, creating a more diverse set of research objectives expressed in a less authoritarian manner. More and more, American archaeology calls for and welcomes multiple perspectives on the past. Today, most archaeologists believe—along with most philosophers of science—that science is part of culture, not outside it. Values, properly factored in, can be productive, not contaminating.

Several chapters in this book draw upon archaeology's postprocessual critique. In Chapter 13, we highlight the research of Thomas Emerson and Timothy Pauketat at Cahokia. Both investigators draw from recently developed postprocessual perspectives to emphasize the importance of ideology and power relations that underlie the hierarchical organization of Cahokia. Our discussions of Poverty Point (Chapter 8) and Serpent Mound (Chapter 9) illustrate how archaeologists are attempting to understand aspects of symbolism and cognition represented in the archaeological record. Chapter 5 emphasizes the multiple perspectives available to understand Wyoming's famous Medicine Wheel.

American archaeologists are today wrestling with several questions: To what extent do we "discover" an objective past? Or are we "creating" alternative pasts from the same data? What is the proper mix of humanism and science in archaeology? What social responsibilities does the archaeologist have to properly use the past in the present? Each of these questions has been around for a century, and no clear-cut resolution appears on the horizon.

The various case studies highlighted in this book were selected, in part, to emphasize the multiple perspectives currently being used to study the past. Although these approaches are sometimes mutually irreconcilable, all have places in modern archaeology because they provide qualitatively different understandings of the human past.

The Modern Business of American Archaeology

Although American archaeology began as a purely "academic" pursuit, without practical application, the huge make-work projects of the Depression set in motion changes that today have revolutionized the practice of archaeology. No longer is American archaeology merely a scholarly exercise conducted mostly by museums and universities. Archaeology today has taken on a more utilitarian role in modern society.

The demands of contemporary life have begun to threaten the common cultural heritage of North America. As transportation systems improve, the past is being paved over. When new houses and hospitals are erected, the debris of the past is carted away in dump trucks. As powerlines go up, sites are bulldozed.

A trade-off is taking place here. Most people would enjoy an improved transportation system, better hospitals, more efficient electrical service. But in the process, America is losing important parts of her cultural heritage. Developers are not evil people, and neither are those wishing to preserve the past. This is a dilemma facing both urbanized nations and developing countries: what part of our past must we save, and what part can we do without?

During the 1970s, the U.S. and Canadian legislatures passed several measures designed to protect and maintain the environment for future generations. The initial charge was to control and lessen air and water pollution, pollution by noise, and pollution by radiation, pesticides, and other toxic agents. In the late 1980s, the mission was expanded to include problems of global warming and environmental change. Significantly, this legislation included archaeological sites as legally protected resources.

So defined, *cultural resources* include physical features, both natural and man-made, associated with human activity. These would include sites, structures, and objects possessing significance in history, architecture, or human development. Cultural properties are unique and nonrenewable resources.

Archaeologists have become involved in *cultural resource management* (CRM). After all, looking out for the global heritage is just one more way of protecting an endangered worldwide environment. CRM in this larger context meshes readily with other efforts designed to comply with current environmental requirements, including legislation protecting endangered species, wetlands, water and air quality, and timber and minerals management.

Obviously, cultural resource management is only a small part of today's concern with saving the planet. And, it must be stressed, archaeology is only part of the CRM efforts (although, in terms of dollars spent, archaeology is a major aspect of the program). Archaeology, in this context, is part of a multidisciplinary effort that draws on the expertise of cultural anthropologists, historians, architectural historians, historical architects, landscape architects, engineers, archivists, and many others to help respect the integrity and relevance of the past. Cultural resource management is concerned with all kinds of historic buildings and structures, such as bridges, artifacts, documents, and, of course, archaeological sites. For decades, cultural resource managers shared in the responsibilities of identifying, evaluating, preserving, managing, and treating these resources.

Archaeological sites located on federal land are legally protected by an elaborate network of federal legislation, policies, and regulations. On state-owned jurisdictions protection varies considerably, ranging somewhere between the firm control exercised over federal lands and the negligible safeguards offered sites on private lands.

CRM is today the most influential force in American archaeology. Increasing numbers of young archaeologists are being hired in the private and government sectors, while the number of more traditional teaching and

9

museum-based jobs has stabilized or even begun to shrink. This means that the bulk of archaeological work being done has shifted from the academic side to the private and government sectors.

It is impossible to tell the story of Native North American archaeology without seriously considering the results of recent cultural resource management projects. In Chapter 10, for instance, I will lead you through the archaeology of Mesa Verde (Colorado), one of America's most important and popular archaeological destinations.

Significant new findings from the Dolores Archaeological Project (DAP) have thrown new light on Mesa Verde. When the McPhee Reservoir and Dam were scheduled for construction near Dolores, Colorado, between 1978 and 1985, federal legislation required that the impacted area be surveyed for archaeological resources. Washington State University played a major role in the DAP, involving dozens of participating archaeologists. They ultimately identified and mapped about 1,600 archaeological sites—including hunting camps, shrines, granaries, households, and villages—in the project area, many of them occupied from A.D. 600 to 900. Roughly 120 of the most critical sites were tested and/or excavated, and the pueblo-style Anasazi Heritage Center was constructed by the Bureau of Reclamation to interpret the research for the broader public. Many of the DAP artifacts are on display there (and the rest are available for study and research).

The Dolores Archaeological Project revolutionized our understanding of Mesa Verde archaeology. No longer can Four Corners archaeology be viewed as a long-term, gradual unfolding of ancestral Pueblo culture. Culture change at Mesa Verde was rapid and far-reaching, and this new perspective is available today only because of the contribution of the Dolores Archaeological Project.

Another example is the important new finds made in recent years as the Arizona Department of Transportation improved the Interstate 10 corridor through Tucson. Three years of work at several different sites along the I–10 right-of-way by Jonathan Mabry and his colleagues at Desert Archaeology, Inc., have radically changed our understanding of the transition to farming-based village life. In Chapter 12, we discuss the implications of these new discoveries for our understanding of Hohokam origins.

The impact of large-scale highway archaeology can also be seen in our discussion of Cahokia, Illinois—the largest pre-Columbian town in North America (Chapter 13). Although archaeologists have worked at this site for more than a century, an entirely new picture of Cahokia emerged as a result of the so-called FAI–270 Archaeological Mitigation Project. The FAI–270 project started in the late 1970s and continued into the early 1990s. Although several institutions participated, the University of Illinois had the longest involvement. This project involved the intensive survey, testing, and broad-scale excavation of numerous Mississippian and pre-Mississippian sites encountered in more than a thousand acres of highway right-of-way.

The FAI–270 project has changed the focus of Cahokian archaeology. The chronology was significantly redefined, particularly for the so-called Emergent Mississippian period. And, for the first time, archaeologists conducted extensive excavations in the Cahokian hinterlands.

In fact, a good-natured competition has arisen between the Dolores and FAI–270 projects for the title of "Largest Archaeological Mitigation Project Ever Conducted in the United States." George Milner, longtime site director of the FAI–270 project, justifies his claim to the title by pointing to the twenty-seven volumes of published site reports and analyses. According to Milner, the FAI–270 volumes takes up a whopping 24.5 inches (62 centimeters) on his bookshelf. William Lipe (Milner's counterpart for the Dolores Archaeological Project) readily admits that the sixteen volumes of excavation reports from the Dolores Archaeological Project take up only 20.7 inches (53 centimeters) of shelf space. But Lipe demands a recount, arguing that once the higher quality paper, smaller type, and microfiche enclosures are factored in, "I am sure the DAP will leave all competitors pitifully far behind."

Regardless of the winner, it is clear that these two massive mitigation projects have demolished traditional views of the past in their respective areas. It is simply impossible today to understand the basics of Native North American archaeology without taking into account the monumental contributions of archaeology conducted as cultural resource management.

On Respecting Native American Viewpoints

In this book, I attempt to provide in clear, jargon-free fashion, a solid scientific baseline of what is known about Native North American archaeology. This is very much a book about ancient stones and bones—the surviving physical evidence left by the first Americans.

But understanding native North America involves more than archaeological excavation and scientific technology. The next federal census will show that more American Indians live in the United States today than when Columbus waded ashore in the Bahamas. The direct descendants of America's first people are still here, and they have a vital stake in understanding their own past and protecting their own heritage.

The fact is, Native American origins are not a mystery to most Indian people. American Indian culture has a cornucopia of oral tradition to explain where they came from. These so-called origin tales address the most profound of human questions: Who are we? Why are we here? What is the purpose of life and death? What is our place in the world, in time, and in space? These questions deal with central issues of value and meaning. Such narrative is so powerful not only because it embodies cultural attitudes, but because origin legends shape cultural attitudes toward fact and reality.

Throughout Native America, stories of human creation reflect a common belief that people are an integral part of the natural world, causally related to the land and trees, the buffalo and the bear. The primordial environment is for many tribes a watery one, with oceans covering the yet-to-be-created earth. Throughout the American West, tribes emphasize the original world parents, Earth and Sky, with ancestral beings arising from the mud to create the earth. For many southwestern groups, several worlds—each with its own colors and symbols—are stacked one atop another; people climbed up through a hole in the ceiling of one dying world into the next, newborn one. Northwest Coast Indian people talk of descending into the modern world through a hole in the sky, much like the smoke hole in a lodge or tipi.

In such societies, where writing and other western devices for "preserving the past" are absent or devalued, this is how historical knowledge is produced and reproduced. And this is how, in societies lacking the services of revisionist historians, history is altered and recast. Long before the emergence of literacy and "history" as an academic discipline, this is how distant events were remembered and imagined. Such creation tales not only construct a workable past, but they also strengthen social traditions and, in the process, create personal identity.

Many Indian people are skeptical of modern scientific explanations. Some reject the chronological tools used by archaeologists to reconstruct the Native American past and many resent the ways that scientists dismiss their long-standing oral traditions as folklore and fairy tales. As American Indians struggle to regain control of their own history, some find common ground with archaeologists. Others don't.

A century ago, archaeologists routinely worked closely with Indian people, and oral tradition was a valid source of information for understanding at least some aspects of the archaeological record, particularly to interpret cultural chronology, site function, and cultural affiliation of the sites being investigated. But during the early twentieth century, archaeologists began to discount the historical value of Native American oral tradition, emphasizing the degree to which Indian tales were "ahistorical," employing as they did nonwestern perceptions of time.

As archaeology became more specialized, direct contacts with Indian people diminished. By the 1970s, few archaeologists had any sustained contact with Native Americans and, not coincidentally, the oral history diminished to the realm of "just-so-stories"—having some ethnographic value for study as such, but providing little factual information about the past or the archaeological record.

I believe that the study of Native American oral tradition can be usefully combined with archaeological evidence to paint a richer, more varied view of the past. In several places throughout this book, I compare Native American oral history with the available archaeological data. The Makah people of the Olympic Peninsula, for instance, have long told traditional stories of massive mudslides that once buried houses, people, and all their possessions at a village named Ozette. In Chapter 6, we see how archaeologists, working closely with the Makah people, have discovered that these oral traditions were true, that buried cedar plank houses—and everything in them—are still perfectly preserved at Ozette.

In Chapter 17, we examine Hidatsa origin tales that describe how ancestral Hidatsa people first came to the Knife River area of North Dakota. One particular Awatixa Hidatsa tradition tells how a supernatural being named "Charred Body" transformed himself into an arrow, and soared down from the clouds to establish the pioneering Hidatsa settlement along the upper Missouri River. Although some American archaeologists dismiss native oral tradition like this out-of-hand, archaeologists working in North Dakota have always recognized the intertwining of Hidatsa oral tradition and their archaeological explorations. We discuss the dramatic evidence at a site called Flaming Arrow, where archaeologists have found startling evidence confirming the Charred Body story of initial settlement by Hidatsa people.

And in the final chapter, we explore the relationship between archaeology and Native American oral tradition for the highly publicized Battle of the Little Bighorn. For more than a century, Native accounts of the battle have been discredited as overly partisan and self-serving fairy tales. Painstaking recent archaeological investigations of the battlefield, however, have conclusively demonstrated that Indian accounts of the battle were considerably more accurate than the *ad hoc* reconstructions pieced together by U.S. military burial details.

In other words, many contemporary archaeologists today are willing to entertain the proposition that real history may be embedded in Native American oral traditions, and that this is the same history archaeologists seek to know from the archaeological record. The question is how to articulate the two sources of information.

Archaeology and oral traditions differ, of course, in how observations are made and interpreted. Western science breaks things down into discrete observational units and measurable variables that can be combined analytically and/or held constant. In general, Native observations and measurements are based on the working of a group of people who view themselves within a holistic environment and societal framework.

The archaeological record is generally cumulative, a compressed record of both past events and the natural and cultural events that have shaped the world. In most cases, the contemporary archaeological record is a modified version of long-term events that have created the present-day landscape. Oral tradition is also cumulative, reflecting the collective remembrance of generations. As time passes, more culturally relevant memories accumulate, while some may be forgotten, discarded, or modified.

These are separate ways of knowing the past, but they tend to converge in a broad sense because certain important issues tend to dominate both realms—migrations, warfare, land use, ethnicity, and so forth. Different standards apply to how relevant information is collected, evaluated, and used. In these pages, we find that both approaches make significant contributions to our knowledge of the Native North American past, but this approach remains somewhat controversial among American archaeologists.

Battling over Bones

Equally controversial is the study of human skeletal remains in American archaeology. Throughout these pages, I highlight contributions from *bioarchaeology*, an important new field that specifically examines the human biological component in the archaeological record. Bioarchaeologists study the origins and distribution of ancient diseases, reconstruct human diets, analyze the evidence for biological stress in archaeological populations, and reconstruct past demographic patterns, all by looking directly at the evidence preserved in human tissues. Also promising is the new and flourishing field of molecular archaeology—itself a subdivision of bioarchaeology—which is taking advantage of new DNA-related technologies to learn about both the very recent and the ancient human past.

When discussing Moundville in Alabama (Chapter 13), for instance, I set out the biosocial implications of the two thousand burials excavated there. In Chapter 13, we consider the evidence for power and authority as dramatically revealed in Mound 72, a corporate mortuary facility designed for the needs of the Cahokian elite. Chapter 14 presents in some detail the contents of the Great Mortuary at the Spiro site—the largest and single richest mortuary deposit ever uncovered in North America. In the final chapter, we see how forensic archaeology has enabled investigators at the Little Bighorn Battlefield to identify and interpret the recently recovered skeletal remains of Mitch Boyer, Custer's mixed-blood Sioux scout and interpreter.

In each case, the bioarchaeological analysis of human skeletal remains contributes significantly to our understanding of the Native North American past. I think these findings are important, and that is why they are included here.

But you should be aware of the current controversies that surround such research. Archaeological scientists have long assumed they had the right to excavate stones and bones to learn about the past. For just as long, Native Americans have questioned this "right," and recent federal legislation has reinforced their objections. Today, archaeologists and American Indians are engaged in a heated and expensive dialogue over the objects of the past and who owns them.

In 1988, the Senate Select Committee on Indian Affairs was told by the American Association of Museums that 43,306 individual Native American skeletons were held in 163 United States museums. Native American representatives pointed out that although Indian people represent less than 1 percent of the U.S. population, their bones comprise more than 54 percent of the skeletal collection in the Smithsonian Institution. Many senators were shocked.

This testimony spurred the U.S. Senate into action and brought an end to decades of wrangling that pitted museums, universities, and federal agencies against Native American tribes. In 1990, Congress passed and President George Bush signed into law a piece of landmark legislation, the Native American Graves Protection and Repatriation Act (NAGPRA).

NAGPRA resonated throughout Indian Country. A significant triumph for Indian people, NAGPRA is important human rights legislation that permits the living to exercise traditional responsibilities toward the dead. NAGPRA also rocked the world of Americanist archaeology, forever changing the business of the past.

NAGPRA covers five basic areas of concern; it

- protects Indian graves on federal and tribal lands,
- recognizes tribal authority over treatment of unmarked graves,
- prohibits the commercial selling of Native dead bodies,
- requires an inventory and repatriation of human remains held by the federal government and institutions that receive federal funding, and
- requires these same institutions to return inappropriately acquired sacred objects and other important communally owned property to Native people.

NAGPRA requires all universities and museums receiving federal money to summarize and inventory such objects. Then, once the items have been identified, the museum community is required to consult with appropriate Native American representatives regarding the "expeditious return" of these funerary objects, sacred materials, items of cultural patrimony, and affiliated human remains.

This legislation has mandated an intensive and continuing interaction between archaeologists and tribal representatives. At first, these interactions were colored with mutual mistrust and apprehension. For decades, many Native American people had felt uncomfortable visiting public museums where their cultural heritage was on display. Some Indian people saw NAGPRA as placing them on equal footing with museum and university officials. Other Native American representatives believed that NAGPRA unfairly favored the museum community, hindering Native people in gaining control over materials that rightfully belong to them (and which never, in their view, should have left Indian land in the first place).

Archaeologists, for their part, are wary of dissolving collections long held in the public trust; such behavior is contrary to every museum charter. Many museum collections contain pieces specifically commissioned for exhibit and for study purposes. Museums would argue that, far from robbing Native people of their heritage, ethnographers and archaeologists have attempted to preserve this heritage for the future and the common good.

At a very basic level, there is complete agreement: all archaeologists, no matter how concerned with preservation of scientific evidence, agree that the bones of known relatives should be returned to demonstrable descendants.

The disagreement comes over defining what is "demonstrable." The specifics of the NAGPRA legislation are still being hammered out.

Despite strong feelings on both sides of the reburial issue, Indian groups and archaeologists are attempting to define some common ground by viewing NAGPRA as a way to foster new collaboration and cooperation in the preservation of Native American traditions and scholarship. A number of repatriations have already taken place, with both sides working toward the common goal of "doing the right thing." A number of tribal museums have already opened, sometimes with the help and cooperation of the museum establishment.

There are some larger issues surrounding first American origins. Can science reserve the right to study matters of the common human heritage? Or does Native American esteem for the dead override scientific rationales? When archaeologists dig up ancient bones, are they serving the greater good or are they robbing the graves of the oppressed? The dialogue over these long-standing questions has recently escalated into a nationwide referendum.

I believe that it is important and appropriate to discuss the new frontiers of bioarchaeological research, and the contents of this book reflect that belief. But I also recognize the need to deal with human remains in a respectful and sensitive manner. Several Native American elders have requested that we refrain from publishing photographs of American Indian human remains. In specific response to this request, no such images appear anywhere in this book.

What's Appropriate when Visiting Indian Country?

Several archaeological sites highlighted in this book exist today on tribal land. Relations between Indian and non-Indian people have varied radically over the years, and the legacy of these interactions is indelible. Many non-Indians are reluctant visitors to Indian Country; others barge right in. The best path is somewhere in the middle.

Indian Country is different from anywhere else in North America. On the "reservations" of the United States and the "reserves" of Canada, you are entering another sovereign country. Keep in mind that tribal authorities have the legal right to control their own natural and cultural resources. Tribal management should be consulted about special regulations and/or permits required for hunting, fishing, hiking, or picnicking. Don't stray from public areas.

For some cultural and religious ceremonies special permission is required to attend, and you should be certain to obtain such authorization. At times, non-Indian visitors are simply not welcome. Sacred sites are particularly sensitive areas, and due to their very nature, few outsiders are welcome there. Local customs vary considerably. It is your responsibility, as a visitor, to find out what is appropriate.

Powwows and certain feast days are festive occasions, and visitors are encouraged to attend. Guests might be allowed to participate in communal

Not long ago, as I was telling my son's third grade class about what it's like to be an archaeologist, a small (but insistent) voice of protest came from the back of the room.

"How come you keep saying 'Indians?' Don't you know they want to be called 'Native Americans?'"

She has a good point. Many people are confused about these terms. In fact, my Native American colleagues tell me this happens to them all the time—people correcting them when they say "Indian"—as if the term had somehow become a dirty word.

Names will always be important, but some names are more important than others. Let's explore the Native American–American Indian issue in a bit more detail.

The word "Indian," of course, is a legacy from fifteenth-century European sailors, who mistakenly believed they had landed in India. The term "American" derives from the first name of another European seaman, Amerigo Vespucci. As Suzan Shown Harjo has pointed out, the term "Vespuccidners" could just as readily be applied to native inhabitants of the land Columbus "discovered."

Many Indian people also point up the ambiguity in the term "Native American." Although I am not an American Indian, I am a native American (because I was born in Oakland, California). The dictionary says that if you're born here, you're a native.

Indigenous people throughout Native North America recognize the garbled logic behind all such labels. Most simply accept the imprecision and use terms such as American Indian, Canadian Native, Native American, Indian, and Native more-or-less interchangeably. We will do the same here.

Of much greater concern to most Indian people is the tribal name. Today, those native Arizonans formerly known as "Pima" and "Papago" prefer to be called the O'odham people. Some Navajo people would like to be known as Diné, a traditional name meaning "The People." Some, but not all, Native people prefer the terms "Lakota" and "Dakota" over the more-common "Sioux" (which is a French variant of an Ojibwe or Chippewa word meaning "enemy"). Whenever discussing a tribe, we will try to use the term preferred by the particular tribe in question.

As Suzan Shown Harjo (Cheyenne/Hodulgee Muscogee) writes in the foreword to *North American Indian Landmarks* (1993), "Do not be discouraged about what is or is not correct, only mindful of what is or is not respectful. The basic rule, as it applies in all human relations, is simply to ask Indians how they would like to be addressed and referred to, and to respect their responses" (xliii).

events such as a "Round Dance," but wait for a specific invitation. Events often proceed according to an internal clock that can disconcert the unsuspecting visitor. Don't plan split-second connections when attending Native American events. Remember that "Indian time" may not necessarily be your time.

Pay attention to the rules (which are only sometimes posted). You will occasionally be asked to leave behind your camera, video equipment, or tape recorder. There may also be prohibitions on smoking, drinking alcoholic beverages, sketching, and/or taking notes. Photography is always a potential point of conflict. Using a flash is usually bad manners at dance ceremonies and contests. A small gratuity is sometimes requested; if you don't want to pay, don't take the picture.

It's generally better to err on the side of formality. Try to be invisible, keeping your questions and interruptions to a minimum. Behave as you would when visiting any other religious service. Respect is often best shown by being unobtrusive. Dress is usually casual, but conservative. Clothing is a form of communication in all cultures. When visiting Indian Country, it's usually a good idea to leave the shorts, halter tops, and other skimpy clothing in the trunk.

Above all, keep your sense of cultural sensitivity close at hand. Forget the racist terms and Indian jokes. Be aware that many Indian people are sensitive about being held up as sports mascots. Leave the Washington Redskins and Atlanta Braves hats at home.

But the key question remains: *Am I welcome in Indian Country?* The answer is almost always "yes." Indian people in recent years are overturning their image as the "invisible Americans." Many tribes have constructed tourist facilities to encourage your visits: hotels, resorts, historical attractions, camping areas, golf courses, recreational facilities, historical attractions, and casinos. Part of the motivation is obviously financial. But many tribes want to show off their heritage and educate you about their past—both the good and bad parts.

If you feel somewhat uncomfortable visiting Indian Country, that's good. Some Indian people feel just as uncomfortable about your being there. But if you're on your best behavior, chances are that your Native American hosts will be too.

Further Reading

An enormous number of books have been written about the archaeology of Native North America. For a general introduction to how North American archaeologists ply their trade, I recommend (of course) my two books on the subject: *Archaeology*, third edition (Ft. Worth, TX: Harcourt Brace, 1998) and *Archaeology: Down to Earth*, second edition (Ft. Worth, TX: Harcourt Brace, 1999). The history of investigation is well-presented by Gordon R. Willey and Jeremy A. Sabloff in *A History of American Archaeology*, third edition (New York: W. H. Freeman and Company, 1993). A very thorough, if slightly dated, overview is Brian Fagan's *Ancient North America: The Archaeology of a Continent*, second edition (London: Thames and Hudson, 1995). I can also recommend *The Smithsonian Book of North American Indians: Before the Coming of Europeans* by Philip Kopper (Washington, DC: Smithsonian Institution Press, 1986) and *America's Ancient Treasures*, fourth edition, by Franklin Folsom and Mary Elting Folsom (Albuquerque: University of New Mexico Press, 1993). Several first-rate entries on North American archaeology appear in *The Oxford Companion to Archaeology*, edited by Brian M. Fagan (New York: Oxford University Press, 1996); *Archaeology of Prehistoric Native North America: An Encyclopedia*, edited by Guy Gibbon (New York: Garland, 1998); and *The Cambridge History of the Native Peoples of the Americas, Volume 1*, edited by Bruce G. Trigger and Wilcomb E. Washburn (Cambridge: Cambridge University Press, 1996).

Blackwater Draw

9300 – 4000 B.C.

Paleoindian and Archaic cultures

in New Mexico

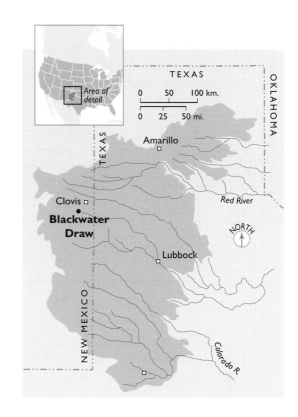

For centuries the world wondered about the Indian tenure in America. Could people have been in the Americas for thousands of years, or were American Indians a relatively recent arrival? The issue was finally resolved in 1927 at Folsom, New Mexico, where human artifacts associated with the bones of long-extinct animals provided unimpeachable evidence that humans—now called Paleoindians—had lived in the New World since the late Pleistocene.

Once archaeologists knew what to look for, new evidence of ancient Americans proliferated. In 1932 another startling discovery occurred about 150 miles (240 kilometers) south of Folsom. A road construction crew digging in a gravel pit near Clovis plowed up a large but extremely well-made stone tool, not far from a huge animal tooth. Archaeologists were notified and important excavations soon began at a site known as Blackwater Draw. Seven decades later, archaeologist George Frison would call it "the most significant Paleoindian site in North America."

The Clovis Discovery

Blackwater Draw is located on the Llano Estacado (Staked Plain), the southernmost extension of the High Plains. The Llano is one of the flattest landscapes on earth, an almost featureless plateau covering

50,000 square miles (130,000 square kilometers). There are no flowing streams, the only permanent water sources being the smallish lake basins called *playas* that commonly cover 5 or 6 square miles (13–16 square kilometers), although they were once much larger.

During the Pleistocene, water streamed through Blackwater Draw, a drainage channel flowing across the western edge of the Llano Estacado. But by the time people arrived here, the climate was already changing, drying the stream into several shallow, seasonal ponds that filled only during periods of runoff. Large mammals—mammoths, bison, and others—were naturally attracted to these ponds, as were their human hunters.

After learning of the finds at Blackwater Draw, archaeologists from the Academy of Natural Sciences of Philadelphia and the University of Pennsylvania Museum began serious work there. E. B. Howard led the charge. He had several things on his mind. The scientific community agreed that people had been in North America for several thousand years before the onset of the Christian era (as it was called then), but Howard saw the need to learn much more about the geological, paleontological, and archaeological sequences of these earliest occupations.

The original gravel pit exposure would eventually become known as Blackwater Draw Locality No. 1. But Blackwater Draw itself extends for miles and contains several "blowouts" of eroded bone and the occasional Paleoindian artifact. Today these other fossil areas have prosaic names like the Oasis Park Locality, the Barrow Pit Locality, and the Model-T (Car Body) Locality.

Howard soon picked up some important stratigraphic relationships in these various exposures. He was well aware of geology's principle of superposition: All else being equal, older deposits tend to be buried beneath younger

Visitors explore Blackwater Draw in 1933, while gravel mining operations continue.

Two huge Clovis points from the Richey-Roberts cache along the Columbia River, near Wenatchee, Washington.

ones. In his regional reconnaissance, Howard observed that the overlying windblown brownish sand contained evidence of more recent pottery-making people. But Folsom-like spearpoints and extinct animal bones were turned up only in the underlying bluish-gray sands, in places where the top sands had been blown away.

Then in 1936–1937 came the real breakthrough. Excavating below the Folsom strata, Howard's team found an unquestionable association between pre-Folsom artifacts and the remains of Columbian mammoths (American elephants). Now Howard and his protégé John Cotter began to search for more subtle differences in the early artifacts they were finding. They soon zeroed in on the characteristically well-made spear points.

Without doubt, several different kinds of projectile points were present at Blackwater Draw, and Howard's team set about trying to separate them. They knew that small, exquisitely made points had been found in some abundance in 1926–1928 at the Folsom site. Terming these "true Folsom points," Howard and Cotter described them as thin, fairly small, and leaf-shaped. Their chief characteristic was a longitudinal groove (the "flute") running along each side or, sometimes, just one face. "True Folsom points" had a concave base, with small earlike projections. The secondary chipping was very fine, showing

E. B. Howard (left) and John Cotter examine Blackwater Draw artifacts in their laboratory at the Philadelphia Academy of Natural Sciences.

remarkable control of the flaking tool. Although these spear points are only about 2 inches (5 centimeters) long, up to 150 minute sharpening flakes were sometimes removed from their surface. In an article published in *American Naturalist* in 1936, Howard called them "the finest examples of the stone-flaking art (319)."

Howard plotted the distribution of "true Folsom" finds, noting that they occurred only in Wyoming, Nebraska, Kansas, Oklahoma, Texas, Colorado, and New Mexico. He traveled to Europe and Russia, studying museum collections from Siberia and elsewhere, searching for prototypes of the distinctive Folsom finds, but he failed to find any. He concluded that if Folsom point technology was not imported, it must have evolved in America. Such an extraordinarily well-made artifact tradition probably did not spring up overnight. The distinctive true Folsom points, Howard argued in the same article, must have been "preceded by other cruder forms which have not yet come to be recognized (319)."

The 1937 excavations at Blackwater Draw clearly confirmed the contemporaneity of humans and mammoths in North America. Because the finds occurred in a deeply buried sand level, there was the promise of establishing a still-earlier sequence of early human artifacts. As Howard had speculated, the mammoth-associated spear points differed in several important respects from true Folsom points, chiefly by being longer and heavier. Howard and Cotter called these larger, earlier, and cruder fluted points "Folsom-like."

Howard suggested an age of perhaps ten thousand years as a conservative estimate for the earliest finds at Blackwater Draw. But when project geologist Ernst Antevs evaluated the geological and paleoclimatological evidence, he expanded the estimate to 12,000 to 13,000 years—a remarkably accurate deduction given the lack of independent dating techniques, such as radiocarbon dating, at the time.

The work of Howard, Cotter, and Antevs at Blackwater Draw stands out as the first truly multidisciplinary attempt to reconstruct late Pleistocene environments in the New World. They not only conducted careful archaeological excavations of the cultural remains, but also initiated detailed stratigraphic and paleoclimatic studies of the sediments and paleontological analysis of the abundant fossil assemblages. They even tried to recover microscopic pollen grains from the ancient sediments.

Refining the Sequence

The Texas Memorial Museum, under the direction of E. H. Sellards and G. L. Evans, conducted excavations at Blackwater Draw in 1949 and 1950, concentrating mostly on clarifying the cultural and stratigraphic sequence of the site. They confirmed conclusively that Columbian mammoth fossils existed only

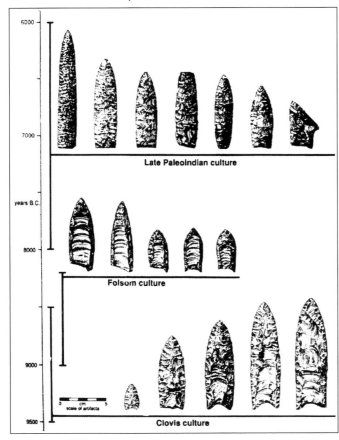

This hypothetical stratigraphic column shows the chronology of various projectile point types during the Paleoindian period.

in the lower fills at Blackwater Draw. They also determined that the Folsom-like points were restricted to the lowest strata at the site. Sellards and Evans dropped the name "Folsom-like" in favor of "Clovis Fluted" and applied the overall term "Llano complex" to the cultural materials recovered from this basal stratum. Above it, in the overlying brown sand stratum, they found true Folsom-style projectile points associated with bones of extinct bison. But mammoth and horse bones were conspicuously absent in this stratum.

Although the stratigraphic details have been refined somewhat since then, Sellards and Evans clearly documented the major significance of the Blackwater Draw deposits: For the first time, it became clear that the elephant-hunting Llano or Clovis complex underlies—and therefore is older than—the bison-hunting Folsom complex. Later still is a third Paleoindian tradition that Sellards called the Portales complex, characterized by well-made but unfluted points. These excavations established the basis of the Paleoindian sequence that archaeologists still use today.

Archaeologists around the world now use the term "Clovis" for the earliest well-documented culture in Native America. In western North America, Clovis sites consistently date between 11,500 and 10,900 B.P.; in eastern North America, fluted point assemblages date slightly later, between 10,600 and 10,200 B.P. Such sites contain thousands of diagnostic artifacts—not only the signature Clovis points but also specialized tools used to process various extinct animal parts.

The Clovis complex provides the earliest well-dated association of human cultural and skeletal remains with extinct animals in North America. The best-known Clovis sites, including Blackwater Draw, are mammoth kills. Bison,

Generalized stratigraphic diagram of Blackwater Draw, showing the position of radiocarbon dated samples with respect to lettered geological strata and cultural affiliations. This stratigraphic section is historically important because it demonstrates unequivocally that Clovis artifacts are older than those of the Folsom culture.

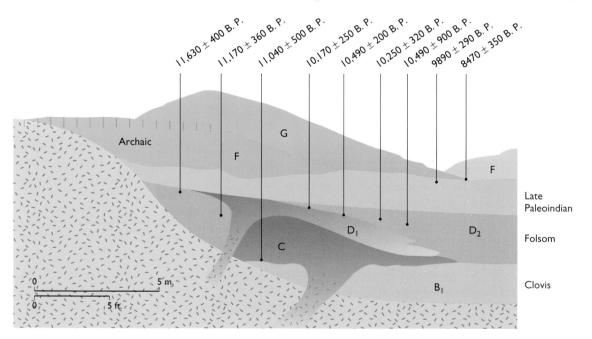

jackrabbit, and birds are also evident at several Clovis sites, including Lehner Ranch and Murray Springs in Arizona. Other foods, such as nuts, seeds, berries, fish, and fowl, were locally available. The Shawnee-Minisink site in Pennsylvania contained seeds from nearly ten species of plants. Investigators have found grinding stones, probably for processing seeds and roots, at several Clovis sites, and some stone knives show a distinctive polish, apparently from harvesting of grasses.

Current Views on Blackwater Draw

Geoarchaeologist C. Vance Haynes has refined the Blackwater Draw stratigraphic sequence. The age of the precultural basal gravel unit (Stratum A) at Blackwater Draw is unknown. These commercial-grade sand and gravel units lie in an ancient stream channel, ultimately deriving from the mountains to the west. They were deposited sometime before the Pecos River captured the Blackwater Draw drainage.

In the depression above the gravels is a fill unit (Stratum B) consisting of gray or "speckled" sand, up to 5 feet (1.5 meters) thick in places, thinning out somewhat toward the margins of the depression. This unit is overlain on the eastern part of the depression by a wedge of springlaid brown sand (Stratum C). The upper part of this unit contains Clovis points in association with mammoth remains, and dates between 11,000 and 12,000 years old. The lower part of Stratum D contains Folsom-style artifacts.

Previous investigators (including Howard and Sellards) believed that Clovis artifacts occurred in the gray sand (Stratum B). But more recent microstratigraphic research demonstrates that the Clovis artifacts had actually worked their way down from Stratum C to Stratum B when that unit was fluidized by later spring activity. Part of the Clovis occupation is therefore assigned to the overlying spring deposits of the brown sand wedge.

The most distinctive Clovis-age fossil is the Columbian mammoth, deposited as both more or less complete skeletons and isolated parts. Also found in Stratum C are remains of horses, camel, bison, turtles, and various small mammals. Two mammoths that Cotter found, now assigned to Stratum C, were heavily butchered with all bones disarticulated, except for part of one vertebral column. One carcass had a fire built on top, suggesting that some of the meat was cooked and eaten immediately after the kill. Clovis artifacts occur in unquestionable association with these mammoth carcasses.

In Paleoindian times, Blackwater Draw was a series of shallow seasonal ponds that collected sediment during runoff periods. Because these ponds attracted large animals such as mammoths and bison, they became prime spots for Clovis hunters who either ambushed watering animals or followed animals wounded elsewhere. Excavations at Blackwater Draw suggest that a camping area existed along the edge of the former pond, with the kills taking place 100 to 200 feet (30 to 60 meters) from the shoreline.

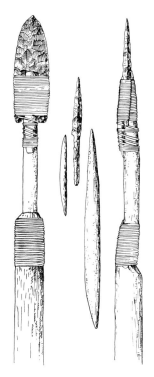

Clovis spear points may have been hafted like this. At the bottom is the thick wooden mainshaft. A smaller bone foreshaft was then attached with sinew wrapping. The stone point may then have been hafted to the foreshaft, held in place with a tapered bone sliver that fit snugly into the characteristic longitudinal flute along the midline of the Clovis point.

The Folsom complex appeared about 10,900 years ago and survived for perhaps six centuries. Mammoths were extinct by Folsom times, and the Blackwater Draw hunters targeted mainly a now-extinct form of bison, as well as occasional pronghorn antelope and mountain sheep. Not as widespread as Clovis remains, Folsom remains are confined to the Great Plains, American Southwest, and the central and southern reaches of the Rocky Mountains.

Paleoindian bison hunters of the North American plains must have been intimately familiar with the landscape and the habits of bison. According to George Frison, they used the best weapons known anywhere in the world at the time. Communal bison hunts were staged with regularity, some Paleoindian sites showing repeated use over several thousand years. Judging from the teeth of the bison killed, communal hunts usually took place in late autumn and winter. The hunters probably froze surplus meat temporarily, then placed it in protective caches to be used as needed.

The overlying Stratum D is made up of diatoms—minute planktonic algae—in a matrix of clay and sand. They were deposited in the quiet waters of a shallow pond that fluctuated between fresh and saline conditions during what was probably a time of dropping water levels, increased evaporation rate, and decreased precipitation. This stratum contains a huge bone bed made up of hundreds of bison skeletons. Neither mammoth nor horse occurs here. Folsom points are found in Stratum D, as are a variety of stone butchering tools.

The dark gray carbonaceous silt of Stratum E contains extinct bison remains and artifacts accompanied by later Paleoindian materials, generally known as Plano (which has replaced Sellards's "Portales complex"). The archaeological evidence indicates that extinct bison were hunted as late as seven thousand years ago.

An erosional hiatus occurs at the top of Stratum E, suggesting a period of extreme aridity between about 5000 and 3000 B.C. Numerous ancient wells were dug at this time, extending 4 to 6 feet (1 to 2 meters) below the ground surface to reach the dropping water table. The ancient well-diggers first cut a circle through the massive clay and diatomaceous earth layer, penetrating the underlying water-bearing sands, which they then easily scooped out to deepen the well. Most of these wells were refilled shortly after use, perhaps to keep them from drying up or being used by enemies. Remarkably similar wells have been found elsewhere on the southern High Plains, suggesting an interval of extremely arid climatic conditions.

Stratum F is a brown eolian (wind-deposited) sand that contains post-Paleoindian archaeological materials from the Archaic period. The uppermost layer, a tan eolian sand called Stratum G, also contains Archaic deposits. This sand-dune sequence shows that the water table has not reached the surface of Blackwater Draw for the past five thousand years. In a climate that probably ranged from warm temperate to hot, Plains Indian groups hunted modern bison here, and they must also have been drawn to the lakes at Blackwater Draw —one of the few water sources on the Llano Estacado. Modern horse bones occur in the uppermost portion, showing that this top stratum persisted into the era of Euroamerican contact.

First American Lifestyles

Blackwater Draw's basic archaeological story is a robust stratigraphic sequence that systematically stacked up the stones and bones of the past. But Clovis people were not stones and bones. They were human beings—the first Americans. We cannot project ourselves back in time and will never encounter a Clovis person firsthand, but we can learn about the past from other, more recent hunting groups, such as the Inuit (Eskimo) caribou hunters and the postcontact bison hunters of the Great Plains. Such knowledge allows us to speculate about what life in Clovis times might have been like.

Clovis men and women must have faced extinction daily. They lived close to the land, and America during the Pleistocene was a tough, unforgiving place. One critical mistake and a hunter could suffer serious injury. If he died, his family was immediately at risk. Clovis hunters competed one-on-one for food with fierce predators and scavengers. Once they acquired food, they had to guard it carefully against this competition.

THE GREAT AMERICAN DIE-OFF

The first Americans witnessed one of the world's most dramatic episodes of extinction. Before their eyes, Clovis hunters saw native animal species die out in droves. The large herbivores were the hardest hit—the 20-foot-long (6-meter-long) ground sloths, giant beavers the size of modern bears, horses, camels, mammoths, mastodons, and musk oxen. As the ecological noose tightened, the carnivores soon followed—the sabertooth cat with its 8-inch (20-centimeter) canines, the American cheetah and lion, and the dire wolf. Perhaps most impressive was the short-faced bear, twice the size of today's grizzly. In North America alone, three dozen mammalian genera disappeared.

Some paleontologists blame the ancestral American Indian for hunting these animals into extinction. Is it mere coincidence, they wonder, that the extinctions took place immediately after Clovis hunters first showed up in the Americas? Perhaps because the large herbivores had never before confronted a two-legged predator, these beasts lacked the necessary defenses, and the Clovis hunters took merciless advantage. Paul Martin's so-called "Overkill Hypothesis" suggests that as Clovis people blitzed their way southward, they carelessly left in their wake the bones of animals rapidly passing into extinction.

Most modern scientists have problems with the Overkill Hypothesis. They emphasize instead the degree to which the Clovis people were themselves at risk during a period of rapid global warming. As the climate changed, sea levels rose, growing seasons became longer, and snowfall and annual precipitation decreased significantly.

Many smaller mammals could adapt to these shifting conditions by modifying their ranges. But the larger ones—the mammoths, mastodons, camels, and horses—placed greater demands on their environments. Unable to cope with their transformed surroundings, they were pushed beyond the brink to extinction.

Maybe human hunters did play a role in wiping out certain animal populations. But most scientists now believe that these animals fell victim to a rapidly changing climate. Clovis people adapted. The extinct megafauna did not.

Despite decades of research, documenting a human presence in the Americas before Clovis has been difficult. Today, a handful of pre-Clovis sites have been documented in North America, the best known being Meadowcroft Shelter (Pennsylvania). This site consists of several occupation surfaces with firepits, stone tools and flintknapping debris, part of a wooden spear, a piece of basketry, and two human bone fragments. On the basis of a sequence of fifty-two radiocarbon dates, archaeologist James Adovasio believes that the Meadowcroft occupation extends (at a minimum) from 14,000–14,500 B.P. to A.D. 1776.

For years, most archaeologists question Adovasio's conclusion. Early stone tools at Meadowcroft are rare and nearly identical to much later artifacts. Characteristic Paleoindian artifacts and Pleistocene megafauna are absent. Although the ice front would have been less than 75 kilometers to the north, the local vegetation around Meadowcroft Shelter seems to have been temperate. But by the late 1990s, corroborative evidence had turned up elsewhere, apparently confirming Meadowcroft Shelter as a viable pre-Clovis occupation.

The most compelling proof of a pre-Clovis occupation comes from Monte Verde, an open-air residential site in the cool temperate rain forests of southern Chile. Tom Dillehay and his colleagues have encountered four distinct zones of buried cultural remains there. Nearly one dozen house foundations and fallen pole-frames of residential huts have been excavated, and fragments of hide (perhaps mastodon) still cling to the poles. Abundant plant remains are associated with the archaeological deposits, as well as numerous shaped stone tools. Because of the muddy matrix, other organic remains have been preserved here, including chunks of meat, wild potatoes, seaweed, and wooden tools. Dillehay believes that the upper layers contain evidence of a human presence between about 12,800 and 12,300 B.P., and most archaeologists accept his conclusion. More controversial are the deeper layers at Monte Verde, which have produced two radiocarbon dates of 33,000 B.P., perhaps associated with stone tools and clay-lined pits.

Monte Verde is critical to our understanding of native North America. Not only does this site break the so-called Clovis barrier, but Monte Verde lies 10,000 miles south of the Bering Straits. Perhaps humans migrated by some route other than the land bridge. Perhaps they left Asia much earlier than previously thought. Or maybe both. Today, the earliest occupation of the Americas appears to be more complex than most twentieth-century archaeologists realized.

Archaeologist George Frison believes that life in earliest America centered on the family. Although capable of great self-sufficiency, Clovis people lived in small informal bands, consisting of perhaps four to ten nuclear families. Political leadership, such as it was, probably fell to a dominant male who derived his authority from well-advertised exploits as hunter and provider. Each band hailed from a traditional territory where men hunted everything but mates. To marry within the band was incestuous.

In times of plenty, Clovis bands gathered together from throughout their broad territories. The elders gambled and exchanged food and gossip. The young people played their own games of skill. They compared adventures and they fell in love.

Hunters spent their lives on familiar ground. Growing up, they discovered the nature and needs of their homeland: how to stalk, where to hide, how the wind worked, how animals behaved when startled. They believed that mammoths and long-horned bison voluntarily made themselves available to humans, but only in exchange for a measure of deference. Disrespect was an affront that not only sabotaged the hunt but also threatened the success of later hunters. Religious specialists were sometimes required to assure appropriate etiquette toward the supernatural.

Reciprocity might have been another survival secret. Regardless of who killed an animal or harvested a plant, everyone was entitled to a share. Even the most esteemed hunter or gatherer of plants would fail sometimes, and the prudent practice of sharing shielded all from short-term setbacks. Great honor was accorded both to those who provided best and to those who shared most willingly. Food hoarding was probably a public and criminal transgression.

Clovis people continued to adapt through the centuries, expanding their range and exploiting the dwindling resources of the open grassland. Then, about 11,000 years ago, the Clovis lifeway became as extinct as the mammoths that the Clovis people hunted. But most archaeologists believe that, one way or another, modern American Indian people are descended from these Clovis pioneers.

Further Reading

The best single source on Paleoindians in North America is *Search for the First Americans* by David J. Meltzer (Washington, DC: Smithsonian Institution Press, 1993). Other important sources include the following: *Clovis: Origins and Adaptations* by Robson Bonnichsen and Karen L. Turmire (Corvallis: Center for the Study of the First Americans, Oregon State University, 1991); *The First Americans: Search and Research* edited by Tom D. Dillehay and David J. Meltzer (Boca Raton, FL: CRS Press, 1991); *Prehistoric Hunters of the High Plains*, second edition, by George C. Frison (San Diego: Academic Press, 1991); *Paleoindian Geoarchaeology of the Southern High Plains* by Vance T. Holliday (Austin: University of Texas Press, 1997); *From Kostenki to Clovis: Upper Paleolithic-Paleo-Indian Adaptations*, edited by Olga Soffer and N. D. Praslov (New York: Plenum Press, 1993); and *Ice Age Hunters of the Rockies*, edited by Dennis J. Stanford and Jane S. Day (Boulder: University of Colorado Press, 1992).

The evidence for Monte Verde is discussed in *Monte Verde: A Late Pleistocene Settlement in Chile. Volume 1. Paleoenvironment and Site Context* (Washington, DC: Smithsonian Institution Press, 1988) and *Volume 2. The Archaeological Context and Interpretation* (Washington, DC: Smithsonian Institution Press, 1997), both by Tom D. Dillehay.

Further Viewing

Blackwater Draw Locality No. 1 (Clovis, NM; 5 miles [8 kilometers] from the south gate of Cannon Air Force Base on SR 467) is open to the public; guided tours are available daily. The nearby Blackwater Draw Museum (Clovis, NM; 12 miles [19 kilometers] south on SR 70) has numerous displays of extinct animals and the ancient weaponry used to hunt them.

TWO

Hidden Cave

3000 B.C.—A.D. 1000

Desert Archaic culture

in Nevada

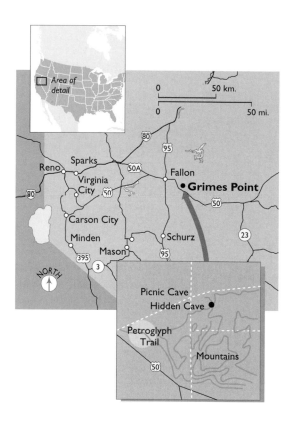

Hidden Cave presides over central Nevada's Carson Desert, which was once a lake. The cave, gouged out about 21,000 years ago by the waves of rising Pleistocene Lake Lahontan, is sealed beneath the cemented surface gravels of the Stillwater Range. The cavern floor was alternatively flooded and exposed until shortly after 10,000 B.P. During relatively brief intervals since then, American Indians crawled into Hidden Cave, leaving behind a well-stratified, well-preserved record of their presence. Natural and cultural deposits continued to accumulate inside until the cave entrance was virtually sealed off by a debris cone. After the cave's rediscovery in the 1920s, teams of archaeologists excavated there in the 1940s, 1950s, and late 1970s.

The dust and darkness inside Hidden Cave created abysmal conditions for Archaic people and archaeologists alike. Because nobody could actually have lived inside the cave, it seems clear that the rich artifact assemblage buried inside must have been deliberately stored away, carefully cached for the future rather than discarded as garbage. Hidden Cave provides important, if unusual, clues about Desert Archaic lifeways.

Rediscovery

"Yup, this is the cave I found back in 1927. No doubt about it."

We had been excavating at Hidden Cave for nearly two summers, and I wanted to believe him. When Dick Wisenhunt appeared at our dig claiming to be the first white man ever to crawl into Hidden Cave, you just had to listen.

His story rang true. Wisenhunt had grown up in Nevada's Carson Desert, on a ranch not far from Hidden Cave (just west of where the Lazy B brothel stands today). A well-known local legend told of a stage holdup in this area. Once caught, the thief confessed to the robbery and claimed that he'd hidden the money in a nearby cave. The holdup story lured four youths, including Dick Wisenhunt, to the cave-riddled hillside. Soon tired of their unsuccessful treasure hunt, the boys started a rock fight. Rocks were flying thick and fast when somebody took cover behind a large boulder. Feeling cold air leaking from the base of the rock, he saw a small dark cleft leading underground. Although wanting to explore further, they were afraid to crawl in "because of the wildcats." So they rocked up the opening and waited months before returning.

Summoning up their courage, the young explorers eventually returned to dig their way into the small cave opening. So difficult was this first entry that

Dick Wisenhunt and unidentified companion at the entrance to Hidden Cave. Wisenhunt claims to have (re)discovered the hidden entrance to the site in 1927 during a rock fight. By the time this photograph was taken in 1933, the entrance had already been considerably enlarged.

Paleoindian lifeways gave way to the Archaic tradition. "Archaic" has assumed a special meaning in American archaeology, referring to the ways people once adapted to the ecology across large parts of Native North America, from the Great Basin, California, and the Pacific Northwest, across the Great Plains to the Northeast and the Deep South. Although the Archaic adaptation spans ten thousand years and a continent, it has certain key characteristics.

When archaeologists refer to Archaic people, they basically mean those hunting-gathering-fishing people who are not Paleoindians. The term was first used during the 1930s for a preceramic, preagricultural culture discovered in New York State. The absence of pottery was considered to be the hallmark of this cultural period. Over the years, as archaeologists expanded their excavations, they found broadly similar materials throughout North America. Today Archaic has two rather different meanings.

The Archaic tradition of eastern North America defines the culture of a specific period: after the initial Paleoindian occupation but before the so-called Woodland cultures, which are generally distinguished by ceramics, mound building, and agriculture. But in western and northern North America, in places where Woodland adaptations did not develop, the term refers to a nonagricultural lifestyle rather than a specific period of time. In California, the Northwest coast, and the Intermountain West, such Archaic lifeways lasted perhaps ten thousand years, well into the period of initial European contact. Chapters 3, 4, 5, and 6 discuss such long-term Archaic adaptations.

There is every reason to believe that Indians of the Archaic period descended directly from Paleoindian ancestors. But the extinction of the Pleistocene megafauna and the spread of the modern deciduous forest produced such significant environmental changes that Archaic people developed rather different lifestyles from their Paleoindian predecessors. As they spread out, they learned to live off the land, and they prospered. Some Archaic groups, particularly those in high latitudes, depended heavily on hunting. Others, such as the Northwest coast groups, became experts at fishing. Some, like the Desert Archaic people of Hidden Cave, relied on seasonal harvests of wild plants, while Native Americans in the eastern Woodlands would eventually discard their Archaic lifeways in favor of farming. Each tradition of the Archaic is adapted to its particular corner of America.

Wisenhunt knew he was the first in recent times to crawl into the crevice. Once inside, they lit up torches—and saw hundreds of ancient Indian artifacts scattered across the cave's huge, flat floor. But the dust and fumes from guano, or bat feces, soon drove them back outside. Pledging one another to secrecy, they once again piled rocks across the entrance. For years, the cave was their secret hideout.

The cave was rediscovered by a guano miner named McRiley sometime in the 1930s, when the substance was in great demand for fertilizer. McRiley wanted to mail a particularly rich sample from the nearby Fallon Post Office but was unable to address the package. He asked the postmaster for help and commented that digging inside the cave would go much quicker "if it weren't for all that Indian junk."

Then, as now, Fallon was a very small town, and archaeologists M. R. Harrington and S. M. Wheeler eventually got wind of McRiley's cave. In 1935 they drove out to take a look. But for all their skittering over the steep hillsides, they couldn't find the rocked-in entrance. After hours of examining promising ledges, an exasperated Harrington allegedly commented, "This is certainly one hidden cave!"—and the name stuck.

Harrington and Wheeler finally located the tiny opening and squirmed in. Blinded by the cave's black interior, they flicked on their flashlights and saw a huge underground room the size of a modern gymnasium. Poking about the guano miner's diggings, they found this hidden cave to be filled with deep stratified deposits, mostly lakeside sediments and slopewash. Ancient artifacts of all descriptions, including basketry, dart shafts with shiny stone points, carved wooden implements, and well-preserved leather fragments, protruded everywhere. The archaeological potential was obvious. Over the next half century, three teams of archaeologists would return to tackle the dust and darkness of Hidden Cave.

Digging at Hidden Cave

In 1940 the Nevada Highway Department sponsored excavations by Sessions M. and Georgia Wheeler. Although the Wheelers were excited by their discoveries, their fieldnotes contain bitter complaints about working conditions inside the pitch-black cave. They lit up the place with carbide lights and tried breathing through a variety of masks and moistened bandanas. But nothing bested the choking dust and darkness.

A decade later, Gordon Grosscup and Norman Linnaeus Roust, two students from the University of California at Berkeley, took up excavations at Hidden Cave. In two months they recovered hundreds of artifacts and excellent samples of coprolites (desiccated human feces), animal bones, and vegetal remains. Knowing the dangers of breathing bat guano, they wore a succession of dust masks, air filters, and moistened cloths. But their 1951 fieldnotes record that "none of these proved satisfactory and until some more capable experimenters produce the answer, the problem will remain annoyingly unsolved."

As it turned out, I was that experimenter. Having guided my students through Hidden Cave in the 1960s and 1970s, I was anxious to see what, if anything, remained unexcavated. Exploring with a weak flashlight, I could see that despite decades of vandalism and illegal relic collecting large sections of intact cave deposits remained untouched.

During the summers of 1978 and 1979, we took another crack at Hidden Cave. Sponsored by the American Museum of Natural History, my team spent weeks trying to solve the logistical difficulties of digging indoors. After installing generators, we experimented with several lighting schemes, finally settling on a combination of fluorescent and quartz-halogen aircraft landing lights. At last, excavators could work in artificial daylight everywhere inside the cave.

But like our predecessors, we found the greatest problem to be the suffocating dust clouds raised by trampling feet. Surgical masks protected our

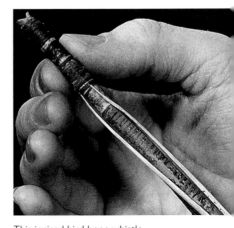

This incised bird bone whistle was cached at Hidden Cave. The delicate sinew wrapping remains intact, and the whistle still sounds a shrill note when blown.

Archaeologists excavating
the lower levels of Hidden
Cave in 1980. To keep from
kicking up dust, they are
standing on wooden walk-
ways. The excavations are
illuminated by quartz halogen
aircraft landing lights, and all
excavators are required to
wear protective dust masks.

lungs, but nobody could see because stirred-up dust particles stayed airborne
for hours. We finally built a network of wooden walkways to keep busy feet
off the fine-grained silts. We also installed a 24-inch (60-centimeter) electric
blower that circulated fresh air throughout Hidden Cave. Although colleagues
ribbed me about "air conditioning" our site, the ventilator kept the cave
sufficiently dust free during working hours.

Who Would Live Inside Hidden Cave?

Digging inside Hidden Cave was no picnic, as three generations of archaeologists can readily attest. But the hassles of working inside this dark dusty
cave taught us some valuable lessons about how the site must have been used
in the ancient past.

Our historical and geological studies confirm that from 2000 B.C. to about
A.D. 1—the period of the site's major human use—the entrance to Hidden
Cave was a narrow tunnel, just like that described by Dick Wisenhunt, and
the interior of the cave was engulfed in disorienting darkness. Breathing
inside Hidden Cave has always been difficult—a condition that would have
been compounded when torches or fires were used for light. For both archaeologists and ancients, Hidden Cave was a very difficult place.

We know from ethnographic studies that Archaic-style people generally
live in places carefully selected to satisfy the minimal conditions of human
life: accessible food, water, firewood, and fresh air; relatively level ground for
working and sleeping; adequate shelter from the elements; and minimally
acceptable levels of heat and light. Hidden Cave comes up short on all counts.

Despite the thousands of artifacts recovered, we were more impressed
with what was not at Hidden Cave: no ash lenses from cooking fires, virtu-

ally no flintknapping debris from making and repairing stone tools, hardly any food leftovers. In fact, the characteristic household debris found in most Archaic-style habitation sites was conspicuously absent from Hidden Cave. Whatever Native American people may have done inside Hidden Cave, they didn't live there.

Tool Caches: What Do They Mean?

Despite the near absence of habitation debris, people had obviously visited Hidden Cave for thousands of years. The archaeological strata are riddled with dozens of ancient caches, storage pits dug into the cave floor. Although most had been emptied of their contents long ago, a few still contained artifacts. These unretrieved valuables tell us a great deal about those who once visited Hidden Cave.

Lewis Binford has made the useful distinction between "active" and "passive" artifacts. An active tool is one that is currently and regularly involved in everyday activities. Active artifacts—manufactured, used, repaired, and eventually discarded—turn up in ancient garbage heaps throughout the world. But tools become passive whenever they are out of synch with daily reality. Our attics and garages contain dozens of passive artifacts—skis last used in February, snow tires removed in the spring, the fly rod ready for opening day, the stadium blanket from last fall's football season, a plastic Christmas tree. Passive gear is only seasonally relevant. During the off-season, it must be stored and cared for, ready to be upgraded to active duty.

These unbroken obsidian (volcanic glass) spear points were made more than 3,500 years ago and cached inside Hidden Cave.

35

So too with the tools of the desert forager. Flat abrasive grinding stones are used to process hulled crops such as piñon (pine nuts). But piñon can be harvested only in the fall, so bulky grinding stones are usually left behind in distant piñon groves, ready for the next fall's harvest. These grinding stones are passive for ten months every year. High desert squirrels and chipmunks hibernate during the winter, so the deadfall snares used to capture them become passive for several months a year. The same is true of fishing gear, duck decoys, and weapons for upland hunting. In general, the more seasonally variable the environment, the more specific the tool kit. The more a group moves around throughout the year, the greater the proportion of the artifact assemblage that passes between active and passive states.

The Hidden Cave excavations turned up thousands of passive, ready-to-go artifacts left behind in well-concealed cache pits. For instance, stone projectile points (dart tips and arrowheads) are common finds in archaeological sites of the American West. Archaeologists usually find fragments, often hunting losses or points broken beyond repair. But at Hidden Cave, more than 80 percent of the projectile points were unbroken and fully serviceable. More than a third had been resharpened in anticipation of future use. These projectile points were not discarded garbage; they were passive artifacts ready to be retrieved when the time was right. Like an attic, Hidden Cave was a safe place to keep valuables secure yet not underfoot.

Food Caches: How Do They Work?

Hidden Cave contained more than passive artifacts. The site was also something like a basement pantry where canned goods, preserves, and other surplus food items are stocked.

Geological cross section through Hidden Cave

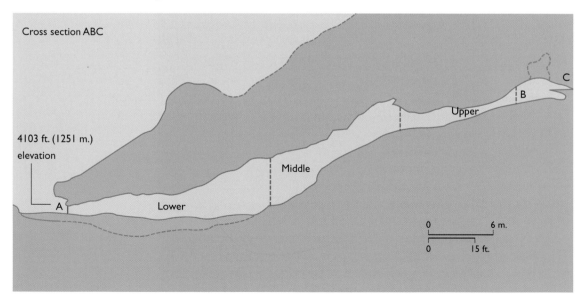

Food caches were part of a much larger survival strategy. We think that these ancient desert people were fairly mobile, timing their movements with the rhythms of the seasons. Fish spawn in the spring, so that is when people remove fishing gear from storage and take it streamside. Hard-shelled seeds ripen in the summer, at which time seed-processing technology is taken to the lowlands. Acorns and piñon nuts are abundant in the fall, requiring carefully timed treks into faraway foothills. These Archaic lifeways played themselves out for thousands of years across the North American desert.

But things rarely meshed that smoothly. What happened, for instance, when all the best resources matured in springtime, with little available during summer and fall? An obvious way to defuse this feast-or-famine problem is to get what you can when things are good, caching what you don't need immediately to use later on. In a way, the food cache is like a tool cache, relegating a temporarily expendable food surplus into a passive state until needed.

Hidden Cave suggests how this strategy may have worked. Intact food caches were relatively rare inside the cave—no great surprise, given their obvious importance for desert survival. But indirect evidence shows that caching food was important to the Hidden Cave visitors.

Everyone who dug at Hidden Cave encountered coprolites, dried human feces. Coprolites are valuable clues about ancient diet—just about the only direct way of knowing what people actually ate. Most of the coprolites found at Hidden Cave contained the undigested remains of a single, simple meal: piñon nut hulls, small fish bones, or maybe bulrush. Our western concept of a "balanced diet" seems to have been of little concern to the people of Hidden Cave. They ate whatever food was immediately at hand.

But one Hidden Cave coprolite contained both cattail pollen and charred bulrush seeds. This might not seem like a big deal, until you remember that cattail pollen is available only in midsummer and mature bulrush fruits can be harvested only six weeks later. This temporal inconsistency means that one or both resources must have been cached for future consumption. Clearly, at least one Archaic-style forager had artificially lengthened the availability of a key resource by storing the surplus away, perhaps inside Hidden Cave, but certainly not far away.

Another coprolite told a rather different story. It contained pieces of piñon nut hull, some bulrush seeds, crushed fish bone, and unidentified seed parts. Once again, this might seem pretty paltry evidence, until you remember what we know about the various plant and animal communities involved. Piñon and bulrush both ripen in the fall, but not in the same place. Bulrushes grow only in marshy desert lowlands, such as the habitat directly downslope from Hidden Cave. But the closest piñon woodland has always been at least several hours' walk—maybe 20 miles (30 kilometers) or so—from Hidden Cave, at a conservative estimate. So this unlovely little coprolite indicates that somebody ate piñon and bulrush on the same day three thousand years ago: a meal that could not have occurred without some fairly serious planning.

37
—

A SECOND HARVEST?

For thousands of years, people stored implements and foodstuffs inside Hidden Cave. But how long does it take to store a few seeds and stash some tools? Presumably not very long. Why then, you might ask, are the Hidden Cave deposits littered with ancient coprolites? Why should so many obey the call of nature, so often, in the few moments they tarried inside the cave?

To complicate matters further, the 1951 excavators excavated what they termed an "aboriginal latrine"—a pit 5 feet (1.5 meters) in diameter, carefully lined with large rocks transported inside, one by one, from the Eetza Mountain hillside. Inside this pit were hundreds, if not thousands, of coprolites. Why would people four thousand years ago use a latrine? Does it really make sense for people to climb halfway up Eetza Mountain to gather the rocks and build a latrine just so they could defecate indoors?

Archaeologist Robert F. Heizer believed that people at Hidden Cave and elsewhere in the Desert West may have deliberately created such latrines as yet another survival strategy—a way of storing undigested seeds for times of extreme famine. The so-called second harvest hypothesis is based on the firsthand accounts of early European explorers, missionaries, and military men in Baja California. In 1771 Father Johann Jakob Baegert in his Observations in Lower California (Berkeley, 1952) described the following experience:

> The pitahayas [cactus] contain a great many small seeds, resembling grains of powder, which for reasons unknown to me are not consumed in the stomach but passed in a undigested state. . . . The Indians collect all excrement during the season of the pitahayas, pick out these seeds from it, roast, grind, and eat them with much joking. This procedure is called by the Spaniards the after or second harvest! . . . It was difficult for me, indeed, to give credit to such reports until I had repeatedly witnessed this procedure. . . . They will not give it up.

"Second harvesting" graphically displays how Archaic-style foragers, at least in Baja, embedded one survival strategy within another. And it may explain the latrine found inside Hidden Cave.

The coprolite evidence demonstrates that storage caches and long-distance transport were deliberate strategies that helped the Archaic-style foragers at Hidden Cave persevere in one of the world's harshest environments.

The Fallacy of the Typical Site

The archaeology of Hidden Cave illustrates a number of strategies that ancient desert dwellers used to survive in their dynamic, if sometimes hostile, environment. But we should not think of Hidden Cave as somehow "typical" of

Desert Archaic sites in general. Taking something as typical—be it an artifact or an entire site—is dangerous, particularly when dealing with the kind of nonsedentary lifestyle followed by many Archaic people.

To see why, look at the diagram illustrating the seasonal rounds of the nineteenth-century Numa (Western Shoshone and Northern Paiute) people, who lived and still live not far from Hidden Cave. Numic people survived by collecting seasonally ripening plant foods, supplemented to some degree by hunting. Because they did not grow crops and were not tethered to agricultural fields, they traveled from one microenvironment to another, harvesting various wild resources.

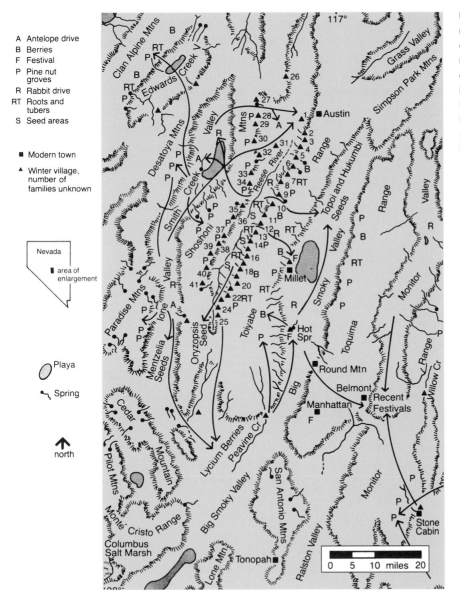

A Antelope drive
B Berries
F Festival
P Pine nut groves
R Rabbit drive
RT Roots and tubers
S Seed areas

■ Modern town
▲ Winter village, number of families unknown

Nevada
■ area of enlargement

◯ Playa
↜ Spring
↑ north

Ethnographer Julian Steward reconstructed this pattern of seasonal movement for nineteenth-century Western Shoshone people living in and around the Reese River Valley in central Nevada, about 100 miles (160 kilometers) east of Hidden Cave.

Nuts of the piñon tree, a Shoshonean staple, ripened in the late fall and often provided enough food for the winter. Piñon nuts are bulky resources, so Numic people generally moved their camps into the foothills to avoid carrying heavy loads across long distances. Buffalo berries and currants also became available in the low foothills about this time. As Indian ricegrass seeds ripened during the early summer, camp was moved from the piñon forest to the flat valley floor. The Numic people used many other local foods in the same cyclical fashion, making critical decisions about whether to move the food to camp or move the camp to food.

Look at the pattern for the Reese River Valley, near the top of the map. The numbered triangles are winter villages, inhabited seasonally to exploit the ripening piñon nuts. The lettered lowland localities were established in the summer for gathering seeds and roots and hunting rabbits and occasionally antelope. Satellite sites served ceremonial purposes; in the upland areas people used them for gathering berries and hunting bighorn sheep. This figure demonstrates only a fraction of the intricate and complex movements and decisions involved in some Archaic-style lifeways.

It also illustrates the fallacy of the typical site. Suppose that you are an archaeologist planning to excavate just one of these Western Shoshone sites. Which one to choose? Winter village sites are attractive because they represent the lengthiest occupation and probably contain remains of a great variety of activities. If you excavated a piñon-gathering camp, you would probably reconstruct a lifeway something like this: "This economy was based on harvesting piñon nuts; the camp contained between one and two dozen people; the men who lived there made lots of stone tools and repaired their weapons; the women spent a great deal of time collecting piñon nuts and grinding them into meal, sewing hide clothing, and making basketry." All these inferences are quite likely, and an enormous amount of ethnographic information about Numic lifeways corroborates them.

But suppose that I decided to excavate a fandango site (or festival, denoted by F on the map). My reconstruction would suggest a grouping of two hundred to three hundred people who subsisted on communal hunting of jackrabbit and antelope, and who spent a great deal of time dancing, gambling, and "living off the fat of the land." And, based on what we know about Numic lifeways, I would be correct as well.

The difficulty is clear: no matter which site is selected, a great deal will be missed. No single Numic site is sufficient to demonstrate the total range of cultural variability. One cannot just dig here or there, because there is no typical site—and Hidden Cave is not typical either.

This twined tule bag had been buried deep inside Hidden Cave. Nineteenth-century Paiute Indians from this area used similar bags when gathering bird eggs.

Further Reading

The best single source for this chapter is *The Archaeology of Hidden Cave, Nevada* by David Hurst Thomas (New York: American Museum of Natural History, 1985). A less technical discussion is "Three Generations of Archaeology at Hidden Cave" *Archaeology* (September/October 1984): 40–47. I also recommend Chapter 13 ("The Great Basin and Western Interior") in *Ancient North America*, second edition, by Brian Fagan (London: Thames & Hudson, 1995). For a general background on the archaeology and natural history of the Intermountain West, see *The Desert's Past: A Natural Prehistory of the Great Basin* by Donald K. Grayson (Washington, DC: Smithsonian Institution Press, 1993).

Further Visiting

Hidden Cave (Fallon, NV; 12 miles (19 kilometers) east on US 50) is today part of the Grimes Point Archaeological Area. The Churchill County Museum and the Bureau of Land Management cosponsor guided expeditions inside Hidden Cave. By the way, you no longer have to crawl inside. Thanks to foresighted Bureau of Land Management engineers, you barely have to bend over. Visitors can also take a self-guided hike along a petroglyph trail outside Hidden Cave.

THREE

Cape Krusenstern

6000 B.C.—20TH CENTURY

Iñupiat Eskimo cultures

in Alaska

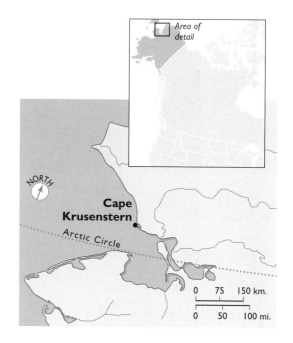

In the late nineteenth century, Edward Nelson made a curious discovery while sailing home after several years of biological and ethnological research on Norton Sound. The trip took him along northern Alaska and part of the Chukotka coast, along the Asian margin of the Bering Strait, where he saw a number of abandoned Native villages entirely unrelated to present shorelines. He thought this odd, as most modern villages were built directly on the shoreline, and he suggested that changing sea levels could account for the presence of these abandoned villages upon relic beach lines.

Archaeologist Henry Collins made similar observations during his fieldwork in the 1930s at Gambell on the northwestern shore of St. Lawrence Island, also in the Bering Strait. As he walked across a series of seventeen sequential beach ridges, Collins believed he was observing a chronological sequence of archaeological cultures: The modern Western Eskimo settlements were built along the modern beachline, and as one moved farther inland, the villages became older. To Collins and other archaeologists of this period, the position of settlements became a rough chronological tool for estimating their age—a way of relating ancient archaeological data to the modern people of the area.

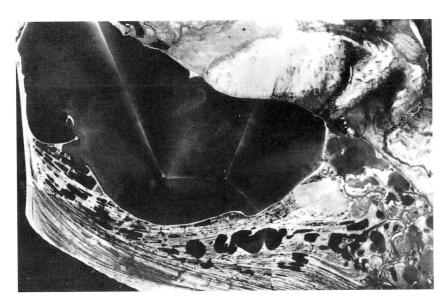

Aerial view of Cape Krusenstern

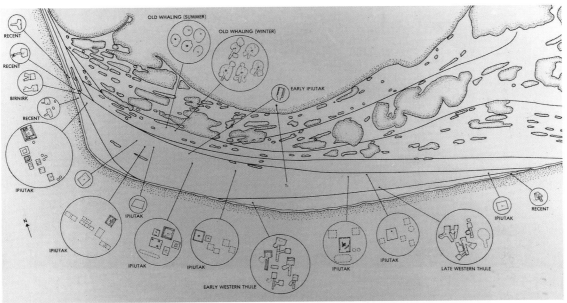

Giddings Arrives at Cape Krusenstern

Location of excavated settlements on Cape Krusenstern

The Alaskan cultural sequence was still poorly understood at mid-century. The issue of chronology brought J. Louis Giddings to western Alaska. Giddings hoped that, because of the excellent preservation of wood in ancient frozen deposits, he could use tree rings to establish a dendrochronological sequence for the Arctic (see Chapter 11 for a more detailed discussion of tree-ring dating). Such a sequence would bring more detailed chronological order to the multiple pre-Eskimo cultures known to have existed there.

What Giddings needed was a series of well-stratified archaeological sites. But vertically stratified sites are rare along the northern Alaskan coast, where most of the archaeology is concentrated instead in single-component shoreline settlements. Even today, no coastal sites north of the Aleutian chain are known to be older than about five thousand years, when the sea reached its present level. Before that time, as the last great continental glaciers melted, sea levels rose significantly, drowning the former coastline and flooding whatever archaeological evidence once existed on the ancient beaches.

So instead of seeking early stratified sites along the northern Alaska coast, Giddings followed Collins's lead by investigating the available shoreline settlements preserved along the northern margin of the Seward Peninsula. Like Collins, Giddings believed that maritime people such as the Eskimo and their predecessors generally built their oceanside villages on the beach ridge with the easiest access to the sea. Therefore settlements on the older beach ridges should predate those on younger beach ridges.

In 1958, Giddings brought his search for ancient beach ridges to the Kotzebue Sound area. He moved his camp to Sisaulik, an early summer village of modern whale and seal hunters. Like many archaeologists, he enlisted the aid of local residents to help find archaeological sites with suitable potential for excavation. Late that summer, Giddings was told about a place called "Sealing Point," known to local reindeer herders for its numerous gravel beach ridges.

Giddings struck it rich at Sealing Point, also known as Cape Krusenstern. The large barren beach is a protracted, relatively flat surface of coarse sand and pea-sized gravel. It slopes upward to a 10-foot (3-meter) high crest not far from the water's edge. Summer storms toss the gravel and sand high up on the beach, well above the zone of normal wave action. As such beach crests stabilize, they sprout isolated stands of grass, followed by a sequence of herbaceous plants that eventually form a dense covering. The row after row of gravel deposits at Cape Krusenstern looked to Giddings "like the parallel furrows of a gigantic field."

Horizontal Stratigraphy at Work

Giddings had stumbled onto archaeology's most dramatic example of horizontal stratigraphy. Aerial photographs show that Cape Krusenstern contains more than a hundred secondary beach ridges that extend far into the Chukchi Sea.

The principle of horizontal stratigraphy is not complex: On any series of uneroded beach surfaces, younger strata will be seaward and older ones inland. The modern Krusenstern shoreline, which Giddings designated Beach 1, contains house pits and other cultural leavings of Eskimo who camped there within the past century. Behind it, as Giddings walked inland from the Chukchi Sea, he counted 113 more beach terraces, most covered by a protective rind of grassy sod. On the mainland behind the Cape Krusenstern beach ridges and lagoon he located sites believed to be even older.

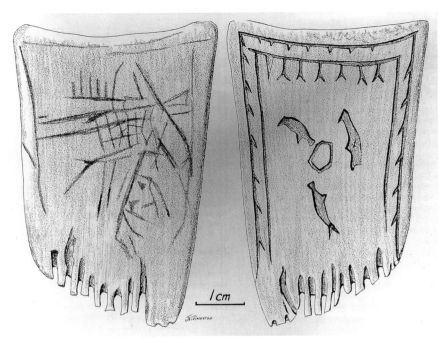

This decorated antler comes from House 35 at Cape Krusenstern, a nearly square, single-roomed house constructed between A.D. 1400 and 1800. The Y-shaped geometric designs appear to be whale fluke representations. The sea mammal carvings probably represent seals.

The Cape Krusenstern archaeological sequence spans at least six to eight thousand years. Not only do these beach ridges document the cultural sequence in this area, but they also provide valuable clues about geological processes. Studies of the ocean sediments indicate that the modern beach is built of gravels that slowly shift southward along the coast, moved by persistent currents. But the beachfront of Krusenstern has switched direction at least six times, changing some 20 to 32 degrees each time. Some geologists attribute this change to shifts in the direction of prevailing winds, coupled with a slight rise in sea levels. Giddings, however, argues that because the earliest archaeological sites have never been washed over by water, sea levels could not have risen more than a yard or so over the past five thousand years.

Defining the Coastal Alaskan Chronology

Giddings began excavating at Cape Krusenstern in 1958 and spent four seasons digging beneath the frozen sod and collecting the ancient beach ridges. He excavated house pits, human burials, artifact caches, tent sites, and entire settlements of those who once lived on Cape Krusenstern. His excitement grew daily as his excavations produced artifacts and house remains unlike anything previously known from Alaska. After Giddings's untimely death in 1964, his student Douglas Anderson continued the research program, eventually establishing a ten-thousand-year sequence for both coastal and interior Alaska.

The archaeology of northern Alaska clearly indicates a continuity of Eskimo culture spanning at least fifteen centuries. In northwestern Alaska, the Northern Maritime tradition encompasses the Birnirk, Western Thule, and

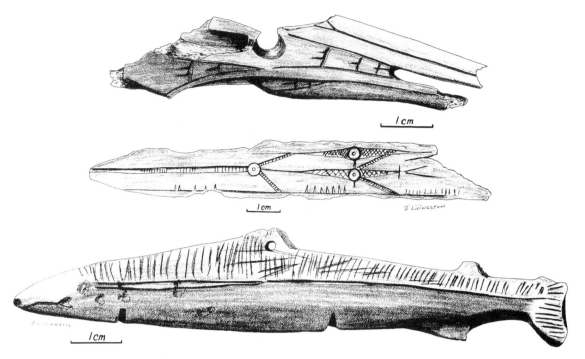

Fish-shaped ivory lure, realistically carved in the round. It was recovered from the late Western Thule House 27, one of the largest and most complex houses excavated at Cape Krusenstern.

all more recent Eskimo phases (A.D. 600–1800). Cape Krusenstern contains more than fifty known house depressions of this period, seventeen of which have been excavated.

Beaches 1–8 at Cape Krusenstern contain the remains of Iñupiat Eskimo camps, clearly evident as square outlines of tent stakes, patterned arrangements of driftwood used as tent weights, and scattered pockets of refuse midden. Nearly thirty house pits dot these forward beaches, some as old as A.D. 1400.

Further inland are remains of the Western Thule culture (A.D. 950–1400), which depended on the exploitation of both sea- and land-based mammals. Cape Krusenstern contains a major winter house with multiple rooms, flanked by single-roomed houses. Such large coastal villages are usually located near whale migration routes, as the Thule people extensively hunted bowhead whales. The highly successful Thule lifeway stretched from northwest Alaska eastward across Canada to Greenland. These Thule people spoke an Inuit Eskimo language.

The earliest manifestation of the Northern Maritime tradition, called Birnirk (A.D. 400–950), is clearly ancestral to recent Iñupiat culture. Birnirk subsistence is oriented primarily to seal hunting but also includes caribou hunting. Birnirk remains are rare in the Kotzebue Sound area and are represented at Cape Krusenstern by a single winter settlement of two or three small houses, each with a single sleeping platform.

Immediately before the development of the Northern Maritime tradition there existed in Arctic America a culture that spanned nearly three thousand years. This was the Arctic Small Tool Tradition, characterized by the miniaturization of various artifact categories; in effect, these people began making composite tools of many small components. The Arctic Small Tool Tradition appeared rather suddenly about 2250 B.C., marking the first colonization of the true Arctic coast. Ranging from Alaska to Greenland, these people may have migrated rapidly from Siberia, where they had once manufactured pottery, but there is little hard evidence for this suggestion. In the eastern Arctic, the Dorset culture appeared about 800–500 B.C. It was characterized by stone houses and perhaps also the domed snow house. People of the Arctic Small Tool Tradition developed an economy that combined sea mammal exploitation with the use of resources from the hinterland.

The Arctic Small Tool Tradition in the Kotzebue Sound area can be divided into five cultural periods: Denbigh Flint (2250–1600 B.C.), early Choris (1650–1250 B.C.), Choris (1200–550 B.C.), Norton-Near Ipiutak (550–50 B.C.), and Ipiutak (A.D. 50–950). The earliest of these, the Denbigh Flint complex, is found at Cape Krusenstern on beaches 80–90. All of these remains represent late spring and early summer seal-hunting campsites. The Denbigh artifact scatters occur as unusually tight clusters, less than 6 feet (2 meters) across. This suggests that Denbigh flintknappers worked around fireplaces within confined areas, such as small tents.

The Old Whaling culture (1400–1300 B.C.) appeared suddenly at Cape Krusenstern, disrupting the long-term occupation of the people of the Arctic Small Tool Tradition. The Old Whalers were well equipped with specialized equipment for hunting whales, whose remains litter the beachline they occu-

Seasonal availability of major animals hunted by Choris people in the Cape Krusenstern

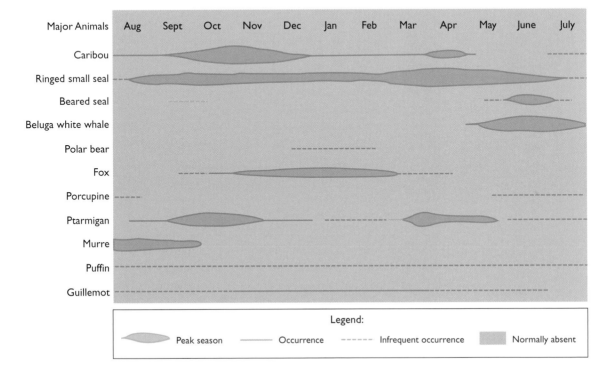

In the popular mind and the Hollywood film, American Indians are associated with bows and arrows. But archaeology tells us that this association has not always existed. Without doubt, the first Americans arrived without the bow and arrow.

The bow appears to have been invented somewhere in the Old World during the late Pleistocene. But nobody knows just where or when. Unmistakable wooden arrow shafts, preserved in archaeological sites in northern Germany, date to about 9000 B.C. The bow and arrow were also in use in Africa at about the same time, or even earlier, judging from the rock art of the area. Preserved bow-and-arrow specimens demonstrate the spread of this technology eastward across Asia, but the trail cools as one moves into northeastern Asia, the presumed origin of the New World populations.

The earliest evidence for bow-and-arrow technology in North America turns up, as one might expect, in the Arctic—perhaps as early as 9000 to 6000 B.C., but certainly by 3000 B.C. The earliest recognizable arrowheads at Cape Krusenstern are in the Arctic Small Tool Tradition assemblages from 2250 B.C. or so.

Whenever it was introduced to the New World, the bow and arrow spread eastward, reaching the Canadian Arctic by 2500 B.C. It moved slowly to the south, reaching the Plains (and perhaps the Great Basin) by about

A reconstruction by John Blitz of the distribution of bow-and-arrow technology into Native North America. Although this reconstruction is probably correct in general, archaeologists still rely quite heavily on educated guesswork in tracking the course of this important invention.

A.D. 200 and spreading into the Pacific Northwest and California by about A.D. 500. Bows and arrows first appear in the so-called Basketmaker Caves of northeastern Arizona about A.D. 500 to 600, and apparently bows and arrows had entirely replaced the earlier atlatl, or spear thrower, in the Southwest by roughly A.D. 750.

How do archaeologists know all of this? The truth is that much of it is inference, even guesswork. Only rarely are archaeologists lucky enough to find the organic remains of bows or arrows in an archaeological site. In the American Southwest, for instance, numerous bow, arrows, and atlatl fragments have survived in arid caves and rock shelters. But although they provide direct evidence of bow-and-arrow technology, the specifics remain rather sketchy because so few of the surviving fragments have been directly dated, such as by radiocarbon dating.

Even such preservation is rare. More commonly, archaeologists must track the spread of bow-and-arrow technology inferentially. Throughout this book I often use the general term "projectile point" instead of "arrowhead," "spear point," or even "bird point" to avoid making assumptions about how a particular artifact was used. But we must make such assumptions when tracking the spread of the bow and arrow. In general, archaeologists believe that the larger points were used on spears and atlatl darts, the smaller ones to tip arrows. Most of what we believe about the spread of bows and arrows through the Americas relies on this inference.

Beautifully carved and decorated ivory snow goggles (or possibly a mask) were recovered from a small hole dug into the floor of Ipiutak House 30 at Cape Krusenstern.

pied. At Cape Krusenstern they are represented by five very large winter houses and five equally large summer houses. In less than a century, however, the Old Whaling culture disappeared.

About 1400 B.C. the Choris culture of the western Arctic introduced pottery north of the Bering Strait, along with oil-burning stone lamps and labrets (lip plugs). At Cape Krusenstern, about two thirds of a mile (one kilometer) inland from the present coastline, the Choris culture left hearths and tent sites, characterized by large spear points remarkably like those used to hunt extinct bison on the western American Plains. By 400 B.C. the rapidly expanding Norton culture had spread throughout the Alaska Peninsula, lasting until about A.D. 1000 south of the Bering Strait. The Norton folk lived along salmon-rich rivers, making pottery, manufacturing ground slate tools, yet still retaining some artifact forms from the Arctic Small Tool Tradition. North of the Bering Strait, Norton was replaced shortly after A.D. 1 by the Ipiutak culture, which was characterized by spectacular ivory carvings but lacked ceramics, oil lamps, ground slate artifacts, and labrets. Beach 35 at Cape Krusenstern has large square pit houses and clusters of shallow summer lodges constructed by the Ipiutak people.

Despite the proliferation of cultural periods, there were overriding continuities within the Arctic Small Tool Tradition, which seem to have contributed (at least in part) to the cultural and genetic makeup of Eskimo people. If so, then the Arctic Small Tool Tradition can appropriately be considered Paleo-Eskimo.

Earlier cultural remains at Cape Krusenstern may be linked to the Arctic Small Tool Tradition, although there is not enough data to demonstrate this. Should this linkage ever be established, then these people may also have contributed, at least in part, to the development of Eskimo culture. The oldest artifacts in the Kotzebue Sound area can be assigned to the Paleoarctic tradition (8000–5000 B.C.), defined by microblades, microcores, flake burins, and large bifaces. These Paleoarctic people hunted mostly caribou and an extinct form of bison and were quite closely related to peoples of northeastern Asia, such as the Siberian Dyuktai culture.

The Northern Archaic tradition, also found at Cape Krusenstern, is

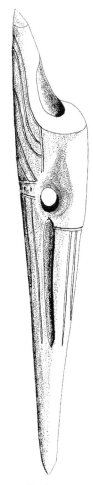

Ivory sealing harpoon head from Ipiutak House 11 at Cape Krusenstern. It is self-pointed, with two inset stone side-blades, and decorated with deeply incised parallel lines from which extend a series of ticked lines.

generally considered to have interior origins, having come to the coast from the northern boreal forests during a climatic warming interval. An ideal caribou hunter's lookout is a site called the Palisades, which sits on a high bench overlooking both Cape Krusenstern and the Chukchi Sea. The Northern Archaic tradition is characterized by side-notched projectile points and scrapers, as well as crude cobble and core tools.

Testing and Extending the Sequence at Onion Portage

Having satisfied himself that beach-ridge dating worked, Giddings attacked the problem of pre-Krusenstern occupations by returning to a more conventional, vertical approach to stratigraphy. He looked for layered deposits deep enough to contain remains of still earlier cultures.

Changing sea levels do not, of course, affect early sites in the Alaskan interior. But these interior sites are also relatively difficult to find. The tundra and taiga offer far fewer resources than the rich shoreline environment, so there are simply fewer sites inland. Because those that do exist can be washed away and buried by shifting river deposits, archaeologists in search of stratified interior sites have concentrated their search on well-drained high ground.

Giddings discovered one such site in 1941, but decades would pass before he realized its full significance. He came across the Onion Portage site while rafting down the Kobuk River, 150 miles (240 kilometers) east of Cape Krusenstern, gathering samples to help establish his Arctic tree-ring sequence. Giddings stopped at Onion Portage to test several house pits and was surprised to find three microcores and a microblade—ancient artifact forms—on a house floor younger than seven hundred years. Over the years, Giddings puzzled over these "anachronistic flints" found among the familiar, more recent artifacts in the Onion Portage house pit.

Later, while piecing together the chronological sequence at Cape Krusenstern, Giddings concluded that microblades had disappeared from the coast 3,600 years ago. These ancient microtools did not belong in a house from A.D. 1400. Maybe the house had been dug into much more ancient sediments. If so, perhaps this site held potential for fleshing out the early end of the sequence. With this possibility in mind, Giddings returned after two decades to the house site on the Kobuk River.

Onion Portage turned out to be the best-stratified site yet discovered in Alaska. For thousands of years, hunters had perched on this sandy knoll, watching massive caribou herds cross the meandering river. People fished nearby and repaired their weapons at campsites littered with stone tools. Over the millennia, spring floodwaters washed over these campsites, leaving thin silt deposits behind, and alluvial deposits washed down from the hillsides, leaving sand strata up to 3 feet (1 meter) thick in places. The result is more than thirty cultural strata in a sequence that is 20 feet (6 meters) deep in places. The bottommost stratum predates the Denbigh Flint complex.

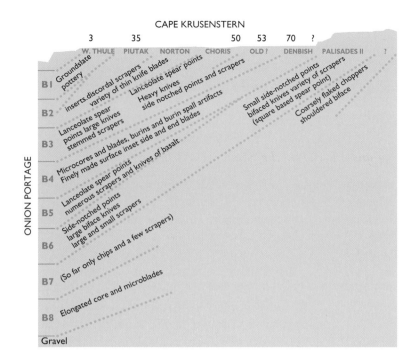

CAPE KRUSENSTERN

3   35   50 53 70 ?

W. THULE   PIUTAK   NORTON   CHORIS   OLD ?   DENBISH   PALISADES II   ?

ONION PORTAGE

B1 Groundslate pottery
inserts discordal scrapers
variety of thin knife blades
Lanceolate spear points
B2 Heavy knives
side notched points and scrapers
Lanceolate spear points large knives
stemmed scrapers
B3 Microcores and blades, burins and burin spall artifacts
Finely made surface inset side and end blades
Small side-notched points
bifaced knives variety of scrapers
(square based spear point)
Coarsely flaked choppers
shouldered biface
B4 Lanceolate spear points
numerous scrapers and knives of basalt
B5 Side-notched points
large biface knives
large and small scrapers
B6 (So far only chips and a few scrapers)
B7 Elongated core and microblades
B8
Gravel

This schematic diagram correlates the horizontal beach stratigraphy at Cape Krusenstern (along the top) with the vertical strata exposed by excavations at Onion Portage.

Radiocarbon evidence shows that the Onion Portage chronology spans at least 8,500 years, perhaps more. Onion Portage not only furnished vertical stratigraphy to supplement the horizontal stratigraphy of Cape Krusenstern, but its deep deposits also effectively bridge the gap between coastal and interior sequences.

Further Reading

The definitive reference for this chapter is J. L. Giddings and D. D. Anderson's *Beach Ridge Archaeology of Cape Krusenstern: Eskimo and Pre-Eskimo Settlements around Kotzebue Sound* (Washington, DC: National Park Service Publications in Archaeology, No. 20, 1986). An excellent, if somewhat dated, overview is *Ancient Men of the Arctic* by J. L. Giddings (New York: Knopf, 1967). A more technical analysis is provided by Douglas D. Anderson in "Prehistory of North Alaska," *Handbook of North American Indians* pp. 80–93 (Washington, DC: Smithsonian Institution Press, 1984). Anderson also wrote an excellent overview of the Onion Portage site in "A Stone Age Campsite at the Gateway of America," *Scientific American* 218, no. 6 (1986): 24–33.

For Further Visiting

Cape Krusenstern National Monument is located on the Chukchi Sea coast approximately 30 miles (48 kilometers) northwest of Kotzebue, Alaska.

Head-Smashed-In Buffalo Jump

3500 B.C.— 19TH CENTURY

Plains Archaic culture

in Alberta, Canada

More than one hundred bison jumps have been identified throughout North America—places where Native American hunters cleverly employed the local landscape to harvest the nearly invincible buffalo that once populated the land. The most famous bison jump is called Head-Smashed-In. Located in southern Alberta, Canada, it is one of the oldest, largest, and best-preserved bison ambush sites in North America. More than 100,000 bison met their death at the cliffs of Head-Smashed-In. Its repeated use over thousands of years is a testament to both the ideally suitable topographic conditions and the skill and daring of the ancient hunters who operated the jump.

Blackfoot people used the site well into the nineteenth century. The old name for these three tribes, Nitsitapii (The People), still applies to them collectively. Today these groups use their own names for themselves: Piikani (North Peigan, pronounced PAY-gan); Kainaa (Many Chiefs Blood, or sometimes just "Blood"); and Siksika (Blackfoot proper). Each used the jump, but the Peigans were (and are) most closely associated with Head-Smashed-In. Their modern reserve is situated directly across the road from the site.

According to Peigan tradition, Head-Smashed-In got its name 150 years ago when a young Indian ventured away from his camp on Willow Creek, wanting to see for himself the dramatic plunge of the buffalo over the steep sandstone cliffs. Standing below the cliffs, like somebody behind a waterfall, he saw hundreds of beasts hurtle to their deaths. But because the hunt that day was so successful, the carcasses piled up in front of him, and he became trapped beneath the cliff. When his people arrived to begin the butchering, they found him with his skull crushed beneath the weight of the dying buffalo. In the Blackfoot language, this place came to be known as *Estipah-Sikikini-Kots*—"where he got his head smashed in."

The Delicate Art of Buffalo Jumping

The Great North American Plains—a flat land of cold winters, hot summers, and sparse and unpredictable precipitation—cover 750,000 square miles (2 million square kilometers). During the Ice Age, this area was a dry, cold steppe grassland, with gallery spruce forests along the main rivers. Here Paleoindians hunted mammoths and other now-extinct Ice Age game. About ten thousand years ago, deciduous trees replaced the conifer gallery forests in the river valleys. In the mountain foothills to the west, Douglas fir and lodgepole pines replaced the open spruce forest of the valleys and slopes. Trees occur today only in stream valleys, scarp lands, and hilly localities.

Plains Archaic people prospered for hundreds of generations by following their natural cycle of hunting game of all kinds and gathering seeds, tubers, nuts, and berries. Although the Plains Archaic lifeway emphasized variety and broad-based subsistence, bison-hunting was always critical for survival.

Early Europeans exploring the Great Plains were astounded by the countless numbers of "wild cows"—American bison—that roamed across an endless "sea of grass." Nobody knows how many bison once populated the Great Plains, but estimates run as high as 60 million animals at the time of European contact. The concerted effort of Euro-Americans to drive the American bison into extinction was nearly successful, with fewer than one thousand animals surviving by the late 1800s.

Long before European horses came to the Plains, Native American hunters developed highly successful ways of harvesting bison. Although sometimes they stalked these huge beasts individually, native hunters knew that driving a stone-tipped arrow or spear through the tough buffalo hide was no easy task. Many arrows were lost before one struck home. This is why the Plains Archaic hunters developed the art of "buffalo jumping," a clever way to take large numbers of buffalo without the dangers and uncertainties of individual stalking. Buffalo jumps employ a highly sophisticated hunting technique, and archaeologists are only now beginning to understand its complexity.

The hunts began spiritually, with medicine men and women carrying out the elaborate and time-honored rituals necessary to insure that the buffalo

The Stampede (1883), by
Frederick A. Verner

A Buffalo Rift, by A. J. Miller

would come close enough to the camp to be easily taken by hunters. This was a deadly serious venture, ultimately controlled by the supernatural. Every Plains tribe had specific songs, charms, dances, offerings, and prayers for calling in the bison. These ceremonies also served to "cloud the mind of the animals," to charm them so they would not realize the existence of the trap. Among the Blackfoot, certain buffalo songs could be sung only during times of near-starvation.

On the night before the buffalo drive, a medicine person would slowly unwrap a pipe and pray to the creator for success. The next morning, the man assigned to call in the buffalo arose very early. He told his wives that they must not leave the lodge, or even look outside, until he returned. They should burn sweet grass and pray for his success and safety. Without eating or drinking, he joined the others and went up on the prairie.

"Buffalo runners" were sent out to locate the herd and begin driving the bison toward the jump. Disguising themselves in buffalo hides and wolf robes, the runners passed near the herds, mostly females, cautiously luring the game toward the cliffs. One specially trained buffalo runner tried to entice the herd to follow him by imitating the bleating of a lost calf. Several days might be required to position the animals for the kill.

The herd was also directed toward the ambush by a carefully positioned set of drive lanes—long lines of individual stone piles, spaced a few yards apart, which served to help speed the stampeding bison toward the cliff. Because bison can only dimly perceive such features on the horizon, they mistook the stone cairns for human hunters and instinctively moved toward the center of the drive lane. The drive lanes sometimes extended for miles, usually forming a V pointing toward the cliff edge.

Once the drive began, dozens (even hundreds) of people hid behind brush piled on the cairns, shouting and waving buffalo hides to keep the animals from turning back. As the herd thundered into the converging drive lanes, hunters ran up from behind, trying to panic the beasts into a headlong rush over the steep cliff. This stampede must have been a frightening sight: buffalo can reach speeds of 35 miles per hour.

The Head-Smashed-In drive lane ended abruptly at an exposed sandstone

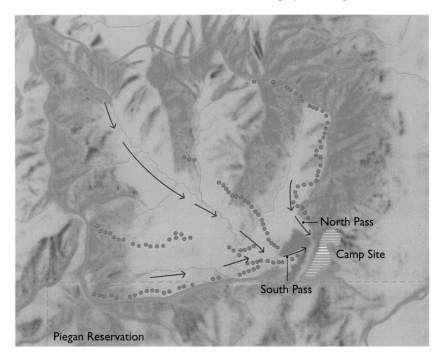

North Pass

Camp Site

South Pass

Piegan Reservation

Plan view of the landscape around Head-Smashed-In. The highland basin covers at least 14 square miles (36 square kilometers), the valleys of the six tributaries of Olsen Creek naturally funneling stampeding bison toward the protruding cliff that towers 65 feet (20 meters) over the talus slope, where the buffalo plunged to their death.

cliff. Because the escarpment faces downwind, bison could not smell the Indians waiting below. As the buffalo reached the cliff edge, most plunged blindly downward. The fall killed many outright; others were disabled with broken legs and backs. A wooden corral may have been built below the cliff, allowing the hunters to easily dispatch the survivors. This step was particularly important because Plains Indians believed that any escaping animals would warn other herds about the jump, and the buffalo would go away.

Archaeological Explorations at Head-Smashed-In

The Head-Smashed-In Bison Jump site, covering 1,470 acres (600 hectares), consists of three major activity areas: the kill site (the jumpoff spot and the bone bed below); the processing campsite on the prairie below the cliff; and the gathering basin, with drive lines marked by stone cairns.

Although George Dawson of the Archaeological Survey of Canada recognized the significance of the site in the 1880s, the first archaeological exploration did not take place until 1938, when Junius Bird of the American Museum of Natural History worked there. Bird mapped the site and made test excavations to a depth of about 13 feet (4 meters).

Boyd Wettlaufer, a pioneer in western Canadian archaeology, was stationed nearby during World War II and returned in 1949 to excavate in the kill site and butchering area. Local ranchers remember the dedicated archaeologist reduced to eating pigeons for weeks after his supplies were exhausted. The Province of Alberta erected a rock cairn to commemorate Wettlaufer's efforts, but their good intentions backfired. The cairn served as a beacon to looters, who shoveled deep into the bone layer at Head-Smashed-In, searching for the exquisitely detailed stone arrowpoints buried there.

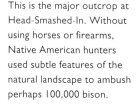
This is the major outcrop at Head-Smashed-In. Without using horses or firearms, Native American hunters used subtle features of the natural landscape to ambush perhaps 100,000 bison.

This illicit looting spurred the Glenbow-Alberta Institute and the University of Calgary to mount a major three-season rescue effort at Head-Smashed-In, producing a great deal of information about the archaeology of this magnificent site. Archaeologists Richard Forbis and Brian Reeves excavated extensively in the kill area, establishing the age of the site and identifying the cultural and technological sequence represented in the 36-foot (12-meter) deep stratified layers of buffalo bones and stone tools.

Although looting continued between excavation seasons, local ranchers spurred on the archaeological efforts, pointing out previously unrecorded drive lanes and even donating land to help establish an on-site public interpretive center.

Archaeologist John W. Brink of the Archaeological Survey of Alberta excavated in the main processing and butchering area from 1983 to 1989. During the 1990s Brian Kooyman of the University of Calgary conducted a four-year project in the North Kill area. The Peigan Reserve terminates only a few hundred yards south of the Head-Smashed-In site, and the Peigan people were closely involved with these archaeological explorations.

In 1981 the United Nations Educational, Scientific, and Cultural Organization (UNESCO) declared Head-Smashed-In a World Heritage Site, kicking off a major effort to develop the archaeological site in a way that would explain northern Plains Indian culture to the broader public. Clearly the intepretive center at Head-Smashed-In would benefit contemporary Native and non-Native people alike.

The Gathering Basin

Surrounded by the Porcupine Hills, the 15-square-mile (40-square-kilometer) Olsen Creek drainage served as the "gathering basin" for the Head-Smashed-In Buffalo Jump. Lush with springs, streams, and luxuriant grass cover, this natural grazing area attracted huge buffalo herds, especially when late-maturing fall grasses were available.

The Indians built a series of drive lanes across the Olsen Creek basin, clearly defining them with more than four thousand small stone cairns (probably once augmented by dung and brush piles that were not preserved). Archaeologists believe that the drive lanes range from five hundred to six thousand years in age. This impressive network of lanes evolved and was improved upon as experience revealed the most successful methods for driving bison. The cairns, known as "Dead Men" by the Blackfoot, were strategically placed to direct the bison herds into narrow valleys or swales leading to the cliff edge; they also served as blinds to hide native hunters. As the herd approached, they stepped out to wave blankets or shoot arrows, moving the herd further into the converging wings of the funnel-shaped drive lanes.

Using this complex feature, the Indians could gather bison from any of several valleys and then drive them for 5 to 8 miles (8 to 13 kilometers). Up

Thousands of stone cairns ("dead men") like this one were constructed to lure bison toward the cliffs at Head-Smashed-In. To catch the buffalo's attention, the rock cairns were probably topped by brush and rawhide strips.

to five hundred people may have been required to operate the Head–Smashed–In gathering basin.

The Kill Site

The kill site consists of a 35-foot (12-meter) high sandstone cliff; at times in the past, it may have been more than 100 feet (30 meters) high. Here the Indians drove the bison to their deaths, creating a huge bone bed at the base of the cliff. In places, the buffalo bones extended to nearly 35 feet (12 meters) below the modern ground surface. The deposits of dirt, stone rubble, bison bones, and stone were eventually covered with windblown dirt (loess) and fallen rock from the overhanging cliff. The bone bed area terminates rather abruptly where Native Americans apparently built a fence or corral along the lower edge of the drop-off area, to contain the accumulating carcasses and prevent them from rolling downhill into the campsite below.

The American bison (*Bison bison*) was the primary game animal for most American Indian groups living on the Great Plains from about 11,000 years ago until late in the nineteenth century, by which time the bison were nearly exterminated.

The buffalo provided Plains Indians with life's needs: food, clothing, and shelter. One bison supplied enough meat to feed an entire tribe for a day. Because a single successful bison drive could easily kill two hundred to three hundred buffalo, the immediate problem became how to process this immense quantity of meat without waste and spoilage. Much of it was sliced into thin strips and hung on racks to dry in the wind and sun. Frequently, this dried meat was then converted to the food staple pemmican. This was done by pulverizing the sundried meat with a stone maul, adding buffalo fat and bone grease and, sometimes, to enhance the flavor, dried chokecherries or other fruit. The mix was placed in a hide container, then pounded to remove all the air. So stored, pemmican could keep for many months. Buffalo tongues were often awarded to those responsible for ensuring the success of the hunt, particularly medicine men and women.

The flesh sustained the hunters, of course, but that was only the beginning. Dressed with the hair left on, buffalo skin protected the wearer from winter's cold; once the hair was removed, it became a summer sheet or blanket or was made into moccasins, robes, and winter blankets. Tanned buffalo skins covered the lodges, the warmest and most comfortable portable shelters ever devised. Using rawhide with the hair shaved off, Plains people made parfleches (trunks in which to pack and transport small items). The tough, thick hide of the bull's neck became a war shield capable of stopping an arrow, turning a lance thrust, or even deflecting the ball from an old-fashioned, smooth-bore gun. The green, untanned hide became a kettle for boiling meat. The skin of the hind leg was made into moccasins and boots. Other parts were fashioned into cradles, gun covers, whips, mittens, quivers, bow cases, and knife-sheaths. Strands of hide were braided into ropes and lines. Buffalo hair was stuffed into cushions, and later saddles. Buffalo horn was fashioned into spoons, ladles, and small dishes. Horns also decorated war bonnets. Glue rendered from the hoofs fastened back and belly sinew to bows to strengthen them. Ribs became scrapers for dressing hides and runners for small sleds

drawn by dogs. Lashed to a wooden handle, the shoulder blades became axes, hoes, and fleshers. Fitted on a stick, the skin of the tail became a fly brush and was used to sprinkle water onto the red hot stones inside the sweat lodge.

Buffalo once grazed in huge herds on the northern Plains; this photograph was taken in the 1880s.

59

Radiocarbon dating suggests that Head-Smashed-In was first used as a buffalo jump about 5,700 years ago. But two spear points found here suggest that the site might even have been used in Paleoindian times (ca. 6500 B.C.), although there is no direct evidence linking the prepared buffalo jump to these artifacts. The partial skeletons of bison suggest that the drives were conducted in the fall, as indicated by the age of the animals killed. For reasons not entirely clear, Head-Smashed-In seems to have been abandoned between about 3100 and 900 B.C.

The most massive and intensive use of Head-Smashed-In took place between A.D. 100 and 850. Most of the butchered bison bones are heavy, non-meaty elements—skull parts, vertebrae, and extremities. The midden contains layers of unburnt, nearly whole bone, interspersed with layers containing only thoroughly fragmented bones. Some strata consist of decayed bone, hair, hide, and manure; other layers contain burnt, unrecognizable bone. Projectile points used to kill or maim the bison are quite common in the midden area, particularly in deposits later than A.D. 150, after the introduction of the bow and arrow.

The most intensively fired strata date to A.D. 1000 to 1200, a pattern repeated at numerous contemporary buffalo jump deposits. This burning may have resulted from massive prairie fires that engulfed the entire region. Associated butchering tools include knives, wedges, choppers, anvils, hammers, bone mashers, and scrapers. Hearths and fire-cracked rock appear in some strata.

The Campsite and Butchering Area

Not far from the cliff and bison jump at Head-Smashed-In is the butchering and processing area. After having quartered the bison carcasses in the kill

John Holloway, a Peigan Indian, excavates at Head-Smashed-In.

area above, the hunters dragged the quarters, hides, and horns to the campsite for further processing. They were probably glad to leave the cramped and putrid killing fields behind, preferring to work on the open flatlands below. A few tipi rings (stones used to anchor the hide tents against the wind) can still be seen on the prairie surface today.

Campsite debris consists of butchered bison bones, bone-boiling and cache pits, cooking hearths, and baking ovens. The artifacts discarded here were used for cooking, smoking, and drying meat; manufacturing stone and bone tools; and processing hides. They consist of pottery, spear and arrow points, knives, scrapers, and a variety of bone tools. The uppermost strata at Head-Smashed-In contain metal arrowheads, knives, and glass trade beads, demonstrating that the jump was used into postcontact times. Once horses and guns became readily available, use of the jump diminished, but did not stop entirely.

Because most of the butchering took place on the open prairie, the deposits are widely spread out, creating a complex patterning of both horizontal and vertical stratification. Some of the most ancient butchering deposits, thousands of years old, are buried to a depth of only 4 to 8 inches (10 to 20 centimeters). In contrast, up at the kill site the butchering deposits stack up, where Reeves and his crew excavated, to a depth of 30 to 40 feet (9 to 12 meters).

In addition to the impressive archaeological record resulting directly from the bison jump itself, the Head-Smashed-In area also contains a number of associated satellite sites. These include burial places, eagle-trapping pits, a vision-quest site, and petroglyphs.

Further Reading

Gordon Reid has written a first-rate, nontechnical summary in *Head-Smashed-In Buffalo Jump* (Erin, Ontario: Boston Mills Press, 1992). George C. Frison puts the Head-Smashed-In Jump in a larger context in *Prehistoric Hunters of the High Plains*, second edition (San Diego: Academic Press, 1991). Brian O. K. Reeves wrote a good, if slightly dated, summary of Head-Smashed-In archaeology in "Six Millenniums of Buffalo Kills," *Scientific American* 249 (1983): 120–135.

Further Viewing

The Head-Smashed-In Buffalo Jump Interpretive Centre (Ft. Macleod, Alberta; 11 miles (18 kilometers) northwest on Hwy. 785, off Hwy. 2) is an award-winning, 10-million-dollar project with exhibits on the ecology, history, and archaeology of one of North America's best-preserved Plains Indian buffalo hunting areas. The Interpretive Centre, built into a grassy hillside, blends into the surrounding landscape. Aboriginal people have played a key role in research and development of the Centre, and highly qualified Native interpreters explain the site and its significance to visitors.

FIVE

Big Horn
Medicine Wheel

18TH–20TH CENTURY

Plains Indian culture

in Wyoming

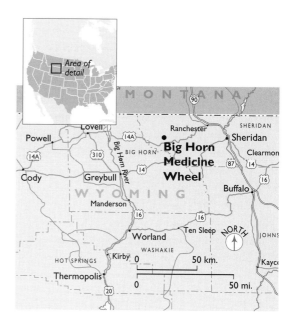

Wyoming's Big Horn Medicine Wheel is simplicity itself: a stone circle nearly 90 feet (30 meters) in diameter. Inside this "wheel" are twenty-eight unevenly spaced stone "spokes," each radiating out from the central "hub," a stone cairn about 15 feet (5 meters) across. Five smaller cairns dot the periphery. The Medicine Wheel was built on an isolated peak 9,640 feet (2,940 meters) above sea level.

Today—as almost certainly in the past—this ancient stone alignment is a sacred site, a place for ritual and religion. Medicine Wheel has long frustrated scientists seeking a precise explanation of its origin and function. Increasingly, archaeologists are turning to a new field, called cognitive archaeology, to make more sense of places like the Medicine Wheel. Although some progress is being made, many archaeologists are a bit uncomfortable with the uncertainty and diversity of opinion that this new approach to ancient ritual and religion implies.

Scientific Explorations

For nearly a century, scientists have puzzled over why anybody would build this high-altitude rock alignment. The western world first took notice of the Medicine Wheel when a 1895 article in *Field and Stream* magazine pointed out certain similarities to Mexico's celebrated Aztec calendar stone.

In 1902, S. C. Simms of Chicago's Field Museum of Natural History was working among the Crow Indians of Montana when he was told of a "medicine wheel" in the Big Horn Range of Wyoming. Simms asked around, but could find no Indians who had seen it firsthand or who knew its exact location. Simms was about to give up when he ran into a white man who had prospected and hunted throughout Wyoming's high country. This man guided Simms to the site. Simms briefly described the rock alignments he saw there, noting that a perfectly bleached buffalo skull had been placed along the eastern side, giving the appearance of looking toward the rising sun. But Simms did not speculate about the uses of the Medicine Wheel, feeling it sufficient merely to call attention to the unusual rock alignment.

Some years later noted western Indian historian George Bird Grinnell visited the Medicine Wheel. His account discusses its general setting and approximate dimensions. Grinnell felt that the wheel must be of great antiquity, and he associated its twenty-eight radii with the twenty-eight roof poles of the Cheyenne Sun Dance Lodge.

In 1958, an avocational group called the Wyoming Archaeological Society

Topographic map of Big Horn Medicine Wheel

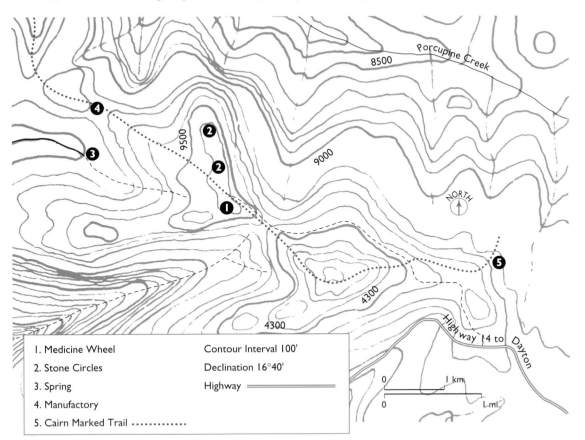

Panorama of the Big Horn Medicine Wheel, taken in 1926. Since then, the stone wall has been removed, and replaced by a chain-link fence.

conducted the first archaeological exploration of the Big Horn Medicine Wheel. They prepared a detailed topographic map of the surrounding area, mapped the visible stone alignments, and excavated two 100-foot-long (30-meter-long) trenches to the north of the Medicine Wheel. They recovered a small quantity of ceramics and glass trade beads, further suggesting a historic-period use of the feature. This team also obtained a single tree-ring date of A.D. 1760 on a piece of wood wedged into the west cairn.

Cairn B at the Big Horn Medicine Wheel, as it appeared in 1902. In the background, a buffalo skull rests upon the east side of the central cairn.

The Wyoming Archaeological Society also excavated the "hub," the central horseshoe-shaped cairn, finding some evidence that a pole had once been inserted into it. Unfortunately, disparities between the final report and an earlier preliminary report cast some doubt on the abilities of an avocational group, however well-intentioned, to interpret the microstratigraphic subtleties involved. The issue is important because one of the primary explanations for the Big Horn Medicine Wheel (discussed below) posits the presence of a central pole, used as an element for astronomical sighting.

Over the decades, several teams of archaeologists have trekked into the Big Horn Mountains to see the Medicine Wheel. Most of these visitors came to the conclusion that the stone alignments were constructed in several stages, because some parts appear older than others. Cultural materials recovered in association with the wheel date to the late precontact and historic periods. There is little controversy on these points.

An Ancient Tomb?

But consensus vanishes when it comes to the questions of precisely who built the Medicine Wheel and why. Some have suggested that the rock cairns were originally constructed as grave markers, each a memorial marking where a particularly powerful person died. The lines of rocks (the "spokes" of the Wheel) show the different directions in which the departed ranged "on the warpath," recording in effect the war deeds of each dead chief. The rock piles at the end of the rock lines may represent enemies killed in battle.

A Vision Quest Site?

Others, drawing upon the rich ethnohistoric record of the Northern Plains, relate the alignment to the ancient practice of the vision quest, a ritual practiced over a wide area in which a solitary individual seeks communication with the spirit world. Although the specifics vary widely, participants are usually sequestered in remote sacred places, without food or water, where they pray for spiritual guidance.

According to a 1926 article by D. W. Greenburg in *The Midwest Review,* the Crow people have explained the Big Horn Medicine Wheel this way:

> Red Plume, a famous Crow chief of the period of Lewis and Clark, obtained his inspiration and received his medicine and the token which resulted in the application of that name by him at the Medicine Wheel. As a young man, Red Plume visited the wheel in the hope of receiving a strong medicine which would make of him a great warrior and chief. Without food, water or clothing, he remained for four days and nights awaiting recognition from the spirits. On the fourth night he was

---

ARCHAEOLOGICAL SITES OR SACRED PLACES?
A NATIVE AMERICAN PERSPECTIVE BY WILLIAM TALLBULL

To the indigenous peoples of North America, the archaeological sites found on North American soil are not "archaeological" sites. They are sites where our relatives lived and carried out their lives.

Sacred sites such as the Medicine Wheel and Medicine Mountain are no different. To Native Americans they are living cultural sites from which help comes when "The People" needed or need help. They are places where tribal peoples went in times of famine and sickness, in periods of long drought when animals would leave, or in more current times when tribes are being torn apart by politics, alcohol, or other abuses.

The men make a pledge to go and vision quest at these places, seeking help. As we leave to go to these sites, our every breath is a prayer. We follow the path to the sites, observing a protocol that has been in place for thousands of years. The Native American approaching these sites must stop four times from the beginning of his or her journey to arrival at the site. A trip to a sacred site was/is not done just for curiosity, but only after much preparation and seeking.

Many blessings have come to "The Peoples" in this way. Many tribes have received covenants (bundles) from these sites. Some tribes still carry the bundles that were received from a certain mountain or site. These are considered no different than the covenants given Moses or the traditional law that went with it.

When Native Peoples have been blessed by a site or area, they go back to give thanks and leave offerings whenever they get a chance. These should be left undisturbed and not handled or tampered with.

Today many of our people are reconnecting with these sites after many years of being denied the privilege of practicing our own religion at these very sacred areas. In the past, trips were made in secret and hidden from curious eyes.

If you go to see a sacred site, remember you are walking on "holy ground," and we ask that you respect our culture and traditions. If you come to a site that is being used for a religious purpose, we hope you will understand.

—The late William Tallbull was an elder, Northern Cheyenne Tribe.

This painted bison skull was situated in a place of honor during the annual Sun Dance. Sometimes the eye and nose cavities were stuffed with grass as an offering to the life-sustaining buffalo. Perhaps the bison skull found at the Big Horn Medicine Wheel served a similar function.

approached by the three little men and one woman who inhabited the underground passage to the wheel and was conducted by them to the underground chamber. He remained there for three days and three nights and was instructed in the arts of warfare and in leading his people. He was told that the Red Eagle would be his powerful medicine and would guide him and be his protector through life. He was told to wear always upon his person as an emblem of his medicine, the soft little feather which grows upon the back above the tail of the eagle. This little red plume gave him his name. Upon his death, after many years of successful warfare and leadership, he instructed his people that his spirit would occupy the shrine at the medicine wheel which is not connected with the rim, except by an extended spoke, and that they might at all times communicate with him there (66).

Red Plume, also known as Long Hair, was a real person, a Crow leader who died in 1836. His visit to the Big Horns could have taken place during the late 1700s.

The Medicine Wheel is, without doubt, well suited for vision questing. In fact, in a 1922 article in *Anthropological Papers of the American Museum of Natural History,* ethnologist Robert Lowie recorded a deposition from Flat-dog, a Crow Indian medicine man who had experienced several visions during his career. Flat-dog describes the Big Horn Medicine Wheel in some detail, explaining that "many of the Crow would go there to fast; the structure has been there as long back as any period alluded to by previous generations. Those who fasted there would sometimes hear steps of someone walking, but looking

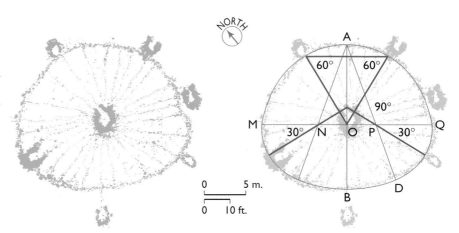

Scale drawings of the Medicine Wheel. The right-hand figure shows a series of geometrical projections superimposed over the Medicine Wheel, demonstrating how the overall shape closely approximates a ring made up of four precise arcs, each of which was probably laid out by stakes and ropes.

up would see nothing. Of such a one the other people were wont to say, 'He is a coward, that is why he did not see a vision.' This meant that he had been terrified by the sound and had looked round, thus losing the vision he would otherwise have secured."

Several contemporary Native American people agree that the site has been used this way. From this lofty perch, the site commands a spectacular view of the Big Horn Basin to the west. Upslope winds whistle through countless crevices, producing moaning and shrieking voices when air velocities are high. Some archaeological evidence suggests that small enclosures protected the rock cairns during the early historic period, perhaps providing a modicum of shelter for those fasting, waiting, and seeking supernatural advice.

An Ancient Observatory?

John A. Eddy set out the most highly publicized hypothesis in the pages of *Science* magazine in 1974. An astronomer by profession, Eddy suggested that Native Americans constructed the Medicine Wheel as an outdoor astronomical observatory. Noting that selected stone cairns might once have held wooden poles, Eddy argued that these posts could have served as foresight and backsight, defining the azimuth of the rising or setting of some important celestial object, probably the sun.

If used for predicting significant celestial events such as the summer and winter solstices, the Medicine Wheel could have imparted powerful knowledge for calendrical, ritual, or even agricultural purposes. Because of its elevation—the site is buried beneath deep snow drifts throughout the winter—Eddy focused on the summer solstice. In a dramatic, televised recreation, Eddy laid out imaginary sighting poles using a surveyor's transit, steel tape, and compass on the morning of 20 June 1972. At dawn on the day of the summer solstice, the sun rose exactly where he predicted it would.

Eddy concluded that simply by observing the sun rising over the cairns at the Medicine Wheel, aboriginal astronomers could have predicted the timing of the summer solstice "with a precision of several days." After further experimentation, Eddy suggested that the other cairns may have marked rising spots of the brightest stars in the summer dawn, stars which themselves further enhanced prediction of the coming solstice.

Skeptics have pointed out several flaws in Eddy's argument. For one thing, the Medicine Wheel lacks convincing evidence for such sighting poles. In addition, the cairns are so large that precise sighting (even with a pole) would not be possible. And even allowing for poles and precision, Eddy's astronomical argument still leaves one cairn unaccounted for.

Archaeoastronomer Victor Mansfield criticizes Eddy's interpretation, suggesting that the "cairns" Eddy employs as backsights and foresights appear more like small huts in which a person might lie down. Mansfield cites an ethnographic account from Flat-head-woman, a Crow who had fasted several times at the Medicine Wheel. According to Flat-head-woman, the virtually naked faster would lie on his back with legs stretched out, the arms extended at the sides, facing east all night. Lowie was told that the faster's "bedding was framed by rocks on both sides."

Other investigators suggest that the Medicine Wheel was built to aid travel, the rock piles left as directional aids to newcomers. Some believe that the Medicine Wheel may have been a boundary marker, a depiction of a stone turtle, or an enduring stone marker demonstrating geometrical expertise. Still others suggest it was built by visiting Aztecs, ancient Moundbuilders,

Eddy's photograph of sunrise (left) and sunset (right) at Big Horn Medicine Wheel. The cairns are in the foreground of each picture.

Russian explorers, gnomes, giants—even miners who wanted to indicate the route to a long-lost gold mine.

The Medicine Wheel as a Modern Sacred Site

After a century of probing and digging, mapping and sighting, the Medicine Wheel is still, as Eddy put it, "a well-known archaeological structure whose origin and purpose remain unexplained." But the Medicine Wheel is much more than an archaeological site. To many contemporary Indian people, it is also a holy place, one of many sacred sites where important ceremonies are performed to this day. Sacred lands are considered to be vital to individual and tribal harmony.

Yet many of the most important sacred sites—places like the Big Horn Medicine Wheel—are overrun each year by thousands of non-Indians: well-meaning tourists, scientific teams, and New Agers seeking spiritual experiences. There is great concern in Indian Country that the plants, paths, shrines, rocks, and other aspects of their sacred sites are being destroyed by the curious and the insensitive.

A coalition representing varied tribal, scientific, ecological, and government interests is working to protect, preserve, and respect the Big Horn Medicine Wheel. In recent years, the Forest Service has closed the last section of access road to vehicular traffic, requiring visitors to travel the remaining 1.5 miles (2.5 kilometers) on foot. This solution minimizes the negative impact of tourism and respects the religious freedom of Native people, yet keeps this powerful place accessible to the many who wish to see firsthand the isolated mountain that has drawn people here for centuries.

Given these diverse opinions and vested interests in the history of the Medicine Wheel, is it any puzzle that so many voices wish to be heard about its meaning? How do you study a place like this? Where is the consensus? Where is the authority? Where is the scientific objectivity? Archaeology's postmodern answer is this: indeterminacy, multivocality, and relativism will probably always thrive in places like the Medicine Wheel. Maybe it's just one of those places that defies finality.

Further Visiting

Big Horn Medicine Wheel (Lovell, WY; 25 miles [40 kilometers] east of Lovell, off US 14A, follow signs) is one of the most compelling Native American religious sites of the past or present. The Forest Service has recently closed the last segment of the road leading to the Medicine Wheel, so be prepared for a 3-mile (5-kilometer) round trip hike. The road is usually free of snow from late June through September.

SIX

Ozette

A.D. I —20TH CENTURY

Northwest Coast Archaic culture

on the Olympic Peninsula, Washington

In 1970, tidal erosion exposed a group of ancient Makah houses in the cliff bank facing the Pacific Ocean. This was Ozette village, where Makah people had lived in their year-round homes into the late nineteenth century. For the next decade, archaeologists and Makahs would work together to uncover the long-term human history of the Olympic Peninsula. But rather than piecing together this past from indiscriminate discards, as usually happens in American archaeology, those working at Ozette had before them the entire range of Makah household items—artifacts from a single moment frozen in time. Ozette is one of archaeology's most significant finds, a true American Pompeii.

The Makah People

The Makah people still live on the northwestern tip of Washington's Olympic Peninsula. In precontact times, they maintained five semiautonomous villages, linked by language, kinship, and common cultural traditions. Each village was inhabited year round, although some people moved to summer residences. Two Makah villages fronted the Strait of Juan de Fuca and three faced westward on the Pacific Ocean. Ozette was the southernmost village, relatively far away from the others.

Ozette village as it appeared in the late nineteenth century. Note particularly the older-style Makah structures, with flat roofs and horizontal, split cedar planking.

Sea and land resources supplied the Makah with a rich subsistence base, which varied somewhat over time and also according to a family's hereditary access rights to hunting and fishing territories. Seal hunting and fishing were open to any able-bodied man. Elk, salmon, halibut, and lingcod were all important in the Makah diet.

For centuries, the Makah lived in huge houses made of split cedar planks. The single-pitched roof, flat but slanted, served as a platform for drying fish and could easily be rearranged for ventilation or removed entirely for use on another house. Paired poles held horizontal wall planks in place. Most houses were occupied by several families, who secured a measure of privacy with temporary partitions that could be removed for dancing and feasting. But because the partitions were only about 3 feet (1 meter) high, people stacked baskets and boxes near them to further separate nuclear family "living areas."

Sustained European contact began in 1792 with the construction of a short-lived Spanish fort. Before long a series of epidemics decimated the Makah population. With the Treaty of Neah Bay in 1855, the Makahs ceded land in exchange for governmental social services and education. The prevailing governmental emphasis on assimilation led to a loss of language and traditional knowledge, and the Indian Service's persistence in introducing an agricultural economy ignored entirely the local expertise in fishing, marine hunting, and navigation.

Throughout the early twentieth century, isolated families continued to live in distant villages, including Ozette. But Neah Bay slowly became the major residential village and economic center of the Makah reservation. Eventually the government required all Indian children to attend a formal school. As there was no school close to Ozette, the seaside village was completely abandoned by the 1930s.

Initial Explorations at Ozette

Richard Daugherty grew up in the forests of the Olympic Peninsula and knew the great archaeological potential of the area. The difficulties of access, however, had prevented much archaeological reconnaissance there. As a graduate student, Daugherty conducted his own archaeological survey in 1947, recording fifty sites, mostly midden accumulations near river mouths. The largest site was Ozette, where huge middens marked a long-term occupation area. Although the occasional Makah family still returned to camp and fish, most of the houses at Ozette had been abandoned a couple of decades before. The site was perfectly situated for whale hunting, and it obviously held great archaeological potential.

Although Daugherty came back from time to time, other archaeological duties distracted him from the Pacific Coast for the next two decades. Then, in 1966, he returned to the Olympic Peninsula as a professor of anthropology at Washington State University to conduct limited test excavations at Ozette. His objectives were straightforward. Using the direct historical approach, Daugherty's plan was to start with the known—the nineteenth- and early twentieth-century Makah occupation—and work toward the unknown, exploring the earlier undocumented occupations at Ozette.

Daugherty enlisted the aid of two colleagues at Washington State University, Carl Gustafson (a specialist in analyzing food bones from archaeological

The stairstep excavations into the Ozette hillside, with the crew from Washington State University. The beachfront strata contain historic period Makah artifacts, but the up-slope deposits are more than two thousand years old.

sites) and Roald Fryxell (a first-rate geologist specializing in archaeological deposits). He laid out an ambitious plan to trench through a series of steplike, wave-cut terraces that rose up to 55 feet (17 meters) above the modern beach-line. Assuming that these terraces reflected the fall of the sea level through time, the earliest occupations should appear on the upper terraces, with later habitation areas clustered near the present beach.

In two field seasons, Daugherty's crew meticulously excavated a 7-foot (2-meter) wide trench extending 230 feet (70 meters) up the Ozette hillside. As expected, the beachfront debris was from the contact period: rusty nails, buttons, buckles, broken ceramic toys, coins, and gun parts. The deposits farther up the slope contained bone fishhooks, stone and shell knife blades, and net sinker stones; some of these deposits were more than two thousands years old. Food bones and shells were abundant, and the investigators took soil samples to be checked for preserved plant pollen.

In their second field season, they extended limited excavations in the historic-period village. Photographs from the 1890s showed a dozen houses built just back from the beach, and Daugherty's team uncovered large quantities of late-nineteenth century debris that the Makah had left behind, including iron ship fittings salvaged from the wreck of the barque *Austria*, which foundered on the nearby beach in 1887.

More than 80,000 individual bones were recovered from the trenching at Ozette, and it became clear that the ancient inhabitants spent more time hunting on the sea than in the forest. Although deer and elk are common on the Olympic Peninsula today, their bones were not abundant in the early excavations at Ozette. Instead, remains of fur seal and sea lions were numerous, and whale bones were so common that they obstructed the excavation trench. After mapping, identifying, and photographing the whale bones, the excavators had to cut them up for removal so that the dig could proceed.

It's a truism in archaeology that the most significant finds seem to occur on the last couple of days in the dig, and Ozette was no exception. The excavators had concentrated on the well-drained, "dry" parts of the site. But they decided to set some test pits into poorly drained, muddy areas to see what might be preserved in waterlogged deposits. The results were promising indeed. The wet clay overburden had, in effect, "sealed" the underlying deposits, and the lack of oxygen had prevented bacteria and fungi from attacking the organic objects buried there.

In one place the excavators found cedar bark rope and mats and several well-preserved baskets. Their progress in a deep test pit was halted by a perfectly preserved cedar plank bench that extended far beyond the confines of the test unit, making it impossible for them to continue. Clearly, part of an ancient house lay buried beneath 10 feet (3 meters) of sand and clay. The carved wooden sleeping bench, mats, and baskets were all preserved. Although only a small part of the house was exposed, Fryxell thought that perhaps a mudslide had destroyed the house, burying and preserving the contents.

Makah traditional stories tell of massive mudslides at Ozette that buried houses, people, and all their possessions beneath tons of oozing wet mud. One account claimed that a village had been buried at Ozette "by a cave-in of the cliff." Mudslides, often triggered by tectonic activity, are not uncommon today along the Olympic Peninsula. In fact, when Fryxell checked survey stakes he had placed at Ozette ten months before, he found them tilting downslope, showing that the slope was still geologically active. If the Makah oral traditions were true, then buried houses and everything in them might still be perfectly preserved at Ozette.

Intriguing stuff . . . but the 1967 field season had drawn to a close. As they broke camp, Daugherty's team could not help but wonder what might still lie buried beneath the slippery slopes at Ozette.

Return to Ozette

Three years later, Daugherty received an urgent telephone message prompting him to rush back to Ozette. The call came from Ed Claplanhoo, chairman of the Makah Tribal Council and one of Daugherty's former students. He reported that a severe winter storm had sent waves slamming into the sea bank at Ozette, causing the saturated bank to erode. Just as Fryxell had predicted, below the mud were well-preserved midden and house remains. Claplanhoo said that looters and collectors were now finding old-style fishhooks, inlaid boxes, and even a canoe paddle washing out from the deeper levels of the site.

After his ten-hour drive to the coast, Daugherty was met by Claplanhoo and a delegation of Makahs. Spilling from the eroded bank at Ozette were the ends of several finished planks belonging to a buried cedar longhouse, a basketry hat, some bone artifacts, and part of a carved wooden box. Daugherty knew that sites mentioned in a 1917 survey of the Washington coast had been completely swept away before his own 1967 survey of the same area. It was clear that Ozette was in danger of destruction, both from looters and continued bank erosion by winter storm tides.

At this point, archaeological and traditional concerns dovetailed perfectly. The eroding cedar plank house was not the one Daugherty's team had uncovered in 1947. It was a second house. Who knew how many more such houses lay buried? Here was an unparalleled chance to document the cultural sequence of this little-known area and an opportunity to find a sample of well-dated, well-preserved, and in-context artifacts—a rarity anywhere in the world.

To the Makah, Ozette represented a way to learn more about their own history. Written records spanned only the last century or so of history as perceived through Anglo-colored glasses. But the Makah people maintained their own rich body of oral tradition, handed down from generation to generation. The finds at Ozette would give them a chance to reconnect with the daily lives of their ancestors, a hands-on way for elders to teach the younger tribal members the lessons of the past. Working with archaeologists from Washington State University, the Makah people could reconstruct their own history as whalers, sealers, fishermen, hunters, craftsmen, and warriors.

The significance of the Ozette project extended far beyond the confines of the Olympic Peninsula. Native peoples were becoming very interested in their own archaeological record—and in what archaeologists were doing with that record. In many parts of North America, confrontations between archaeologists and Indian people were uncompromising and unpleasant. Some still are. But at Ozette all parties agreed that the site contained critical information and required extensive, immediate archaeological investigation.

Knowing that time was short, the Makah Tribal Council joined Daugherty in appealing directly to U.S. Representative Julia Butler Hansen and Senator Henry M. Jackson for help. Jackson quickly arranged for $70,000 to be transferred from the Bureau of Indian Affairs so that work could begin immediately to re-open the excavations at Ozette in 1970. The project eventually blossomed into a year-round endeavor that continued until fieldwork closed down in 1981.

An existing supply trail, running 3.5 miles (6 kilometers) through the rain forest, was upgraded to a "boardwalk" to accommodate heavy visitor use during the main excavations. After following this boardwalk visitors emerged from the forest and walked for half a mile up the beach. More than 25,000 of them toured the Ozette excavations each summer in the middle 1970s—a remarkable figure considering unpredictable coastal weather of the temperate rain forest and the 8-mile (13-kilometer) trek involved.

Wet-site Digging

Most of Ozette is a standard "dry" archaeological site—an extensive shell midden deposit spanning the last two thousand years. Like dozens of similar places along the Olympic Peninsula, Ozette contains large quantities of non-perishable food refuse, ash, and cultural materials deposited along the sandy Pacific beach terraces.

But in archaeological parlance, Ozette is also a "wet site," meaning that the deposits are so water-saturated that destruction by fungi and bacteria has been radically reduced. Finding a "wet site" is phenomenally good luck. Instead of recovering merely nonperishable stone and bone remnants, investigators may also find fragile materials made of wood and fiber. The wet part of Ozette preserves a cultural record without parallel on the Northwest coast. About three hundred years ago a destructive mudflow flattened part of the village, crushing houses and burying thousands of artifacts beneath a 10-foot (3-meter) deep mantle of blue-gray clay. What Pompeii had preserved by volcanic ash, Ozette protected with mud.

Traditional "dry land" excavation methods were worthless for the Ozette house deposits. The preserved basketry and wooden artifacts were much too

This excavator is using a high-pressure firehose to clear away clay overburden from the upper part of House 1 at Ozette. Note particularly the rafter-support post in the foreground.

Once the Ozette house remains were exposed, they were mapped, element by element, prior to removal and stabilization.

delicate for shovelwork. Even the archaeological standard, the Marshalltown trowel, cut through fragile artifacts before they could be seen. But the trick to excavating a waterlogged deposit is simple: add more water. Archaeologists working at Ozette perfected the techniques of hydraulic excavation. Coast Guard and Marine Air Reserve units helped by airlifting in four Briggs and

Stratton-type pumps, which sucked in seawater through intake hoses strung across the beach to a large tide pool several hundred yards west of the site. Despite occasional clogging with seaweed and jellyfish, the system worked well, delivering a steady supply of high-pressure seawater to the excavations above. Water pressure could also be decreased so that excavators could select exceedingly fine sprays to expose fragile basketry, blankets, cordage, and even still-green tree leaves.

There was, however, a trade-off. Because Daugherty had spent two seasons carefully excavating the "dry," midden portion of the site, he decided to sacrifice the strata overlying the buried houses. So, using hoses capable of delivering 250 pounds of water pressure, he hydraulically stripped away the upper layers as rapidly as possible down to the heavy debris flow of the mudslide. It is always a difficult decision to trash part of an archaeological site to expose what lies below, but getting to the newly exposed house at Ozette was an urgent matter. Waves were already washing it away.

Daugherty's team eventually exposed several houses at Ozette, and they were huge. The largest excavated house measured 67 by 39 feet (20 by 12 meters); a second house was about 56 feet (17 meters) long by 31 feet (9.5 meters) wide. Although these houses tend to be smallish by historic Northwest coast standards, the houses at Ozette are still about the size of a modern tennis court.

Plan view of Houses 1 and 5 at Ozette village.

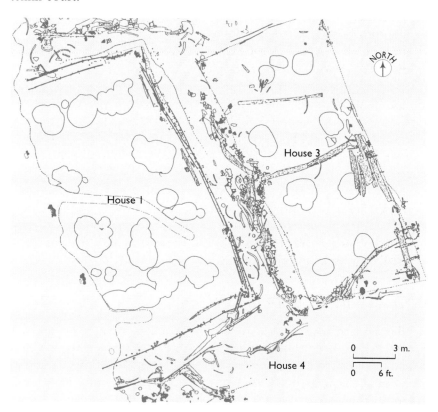

Nearly 20,000 structural members were recovered from the excavated houses at Ozette. These included planks for roofing and walls, 3-foot (1-meter) wide sleeping benches, drains, and rafter support posts. At least three of the houses contained drainage-control trenches or drains, essentially storm sewers that diverted ground runoff from the house floors. In fact, during the water excavation of House 1, the then-undiscovered "under-floor" drain in House 2 continued to operate efficiently, routing excavation water runoff downslope while remaining unclogged. When House 2 was finally uncovered, excavators marveled at the efficiency of the three hundred-year-old engineering. The Ozette villagers had also known how to stabilize the steep hillside on which they lived, using skillful landscaping that long prevented downslope movement.

The artifact inventory from Ozette is truly phenomenal, numbering in excess of 50,000 items. There were woven cedar bark pouches containing harpoon blades, wooden clubs for dispatching seals and fish, fine tools for wood-carving, pendants, hats, baskets, toys, weaving equipment, and an extraordinary life-sized wooden effigy of the dorsal fin and back of a killer whale, studded with more than seven hundred sea otter teeth.

Uncovering and mapping the waterlogged remains was only the beginning of analysis and stabilization efforts. Once the recovered wood and fiber objects had been exposed to air, the decomposition process began immediately. If left untreated, the waterlogged organic artifacts from Ozette would quickly turn brittle and crumble. After considerable experimentation, the Ozette team elected to use a preserving solution of polyethylene glycol, commonly known as Carbowax. This pasty wax, melted and diluted with water, filled large soaking tanks located on-site and in the Neah Bay laboratory. There, the fragile baskets, wooden boxes and bowls, and smaller building elements were saturated with Carbowax. The huge cedar structural remains posed a particular problem. After being exposed, mapped, and photographed, they were carefully hoisted to on-site preservation tanks, where they were soaked in preservative. Conserving this huge artifact assemblage has been an extremely time-consuming process; some of the artifacts require months or even years to become completely saturated. The hardwoods are particularly difficult. Their pores are too small for Carbowax to penetrate, but oxygen can get into these tiny spaces, causing the artifact to rot from the inside out when exposed to air for long periods of time.

What about the Mudslide?

No European trade items, such as glass beads, turned up in the houses. The layer of cultural materials associated with the destroyed houses has been radiocarbon dated between A.D. 1415 and 1621. One of the houses has a preliminary dendrochronology date of A.D. 1719; because the outer rings were missing on this sample, this tree-ring date means only that the mudslide must

Richard Daugherty with a remarkably well-preserved cedar whale effigy, inlaid with more than seven hundred sea otter teeth. The teeth set along the base form a design of a mythical bird carrying a whale in its talons.

have taken place sometime after A.D. 1719 (for a discussion of tree-ring dating, see Chapter 11). Taken together, this evidence suggests that the slide took place roughly three hundred years ago, not long before the earliest face-to-face European contact.

An intriguing alternative has recently surfaced. Recent research suggests that the mudslide might have been triggered by a tsunami (a tidal wave caused by an earthquake) in 1700 whose existence was recently rediscovered in Japanese historical records. If so, then the catastrophe at Ozette might have taken place in January 1700.

Whatever the exact date, it is clear that the mudslide hit part of the village with freight-train speed and force. Massive rafter support posts were snapped and house tops were sheared off. What remained was buried by 6 to 10 feet (2 to 3 meters) of clay.

Ozette Lifestyles

The archaeological record of Washington's coastline gives an impression of long-term equilibrium and continuity. Little at Ozette contradicts knowledge of ethnographic patterns on the Olympic peninsula, but the wet context offers a huge potential for examining in detail coastal life before the arrival of Europeans.

Ozette was a large multiseason community. During the fall and winter months, everyone lived there, and the village population peaked. As elsewhere on the Northwest coast, winter was a time for gathering together and feasting. Then, during the spring and summer, the population dispersed as able-bodied villagers moved out to live in distant camps or villages where they could hunt and fish close to their temporary camps. During the summer, fewer families remained at Ozette to fish for halibut and sockeye salmon.

Whaling was clearly the most important single subsistence activity at Ozette, accounting for three-quarters of the total meat and oil represented by the faunal remains. It is difficult to imagine a more thrilling and dangerous pursuit. Eight men handled the cedar canoe, the "whaler" standing on the prow, ready to thrust his 80-pound (35-kilogram), shell-bladed harpoon into the 30-ton whale, which would sometimes breach within arm's-length of the canoe. At any moment, the fluke could shatter the canoe into kindling, dumping the crew into the frigid Pacific—which meant death in minutes unless the shoreline was close at hand. The Makah and their Nuu-chah-nulth (Nootkan) neighbors claim to be the only people south of Alaska to actively hunt whales on the open ocean, venturing up to 20 miles (32 kilometers) from shore.

This infant's face covering from Ozette is one of the 50,000 items found at the site.

During the nineteenth century, being a "whaler" (harpooner) was a hereditary and prestigious occupation, although crew members were selected at least in part for their hunting ability as well as their family history. At Ozette, the distribution of decorative materials in households clearly demonstrates differences in social standing between families. Distributional differences in whaling trophies and other symbolic paraphernalia demonstrate differences in economic and social standing, while subtle differences in the kinds and proportions of food items among the various houses further suggest that some lineages enjoyed access to the better resource areas, while other hereditary areas were less favored.

The families at Ozette maintained exchange networks with far-flung neighbors. Their houses contained dentalium and abalone shell from western Vancouver island, coiled basketry from the Puget Sound–Georgia Strait region, and carved wooden bowls from the mouth of the Columbia River.

MAKAH PERSPECTIVES ON ARCHAEOLOGY AT OZETTE

We Makahs look in a special way at what is coming from the mud at Ozette, for this is our heritage. Now we can see what life used to be like and add this to what we remember and what our elders have told us. We can hold the household equipment and the hunting and fishing gear that our ancestors designed and made and used. The carving on the boxes and bowls and hunting clubs shows that our people have always valued beauty. The harpoons of yew wood with mussel-shell blades and elk-bone barbs used for taking whales show the strength and skill and bravery of the people.

Our ancestors lived richly by utilizing the raw materials of the forest and the sea. A few of us can remember the very last of those old days. All of us have heard our parents and grandparents tell about them. Now our young people can see for themselves, and so can all the nation. Ozette is like opening the door of a museum—for us, a museum of our own past.

—The Makah Tribal Council
(from Ruth Kirk with Richard D. Daugherty, *Hunters of the Whale*, 1974)

This bowl for oil was carved into a male form, with exquisite detail, even down to the braided hair.

The spectacular archaeological finds recovered at Ozette sparked an interest in Makah language and culture, causing the tribe to commit significant resources to the development of a tribal museum to display and maintain the important Ozette archaeological collection. A longhouse was built in Neah Bay combining traditional knowledge with construction details garnered from the excavations at Ozette. Not only has the Makah tribe placed the priceless articles from Ozette on public display, but the Makah Cultural and Research Center, opened in 1979, continues to make the collection available for archaeological research.

A number of young Makahs worked with the archaeological teams at Ozette, and several subsequently received college degrees in anthropology. The Makah Cultural and Research Center today administers a highly successful language preservation program in which young Makah learn to speak their ancestral language and to create a modern written equivalent.

Further Reading

The best nontechnical discussion of the Ozette site is *Hunters of the Whale* by Ruth Kirk with Richard D. Daugherty (New York: William Morrow, 1974). I also recommend "Ancient Indian Village Where Time Stood Still" by Richard D. Daugherty and Ruth Kirk, *Smithsonian Magazine* 7, no. 2 (1976): 68–75, and "Ozette" by Marcia Parker Pascua, *National Geographic* 180, no. 4 (1991): 38–53. Important archaeological summaries have been published in *Ozette Archaeological Project Research Reports, vol. I. House Structure and Floor Midden* (Pullman, WA: Washington State University and National Park Service, Pacific Northwest Regional Office, 1991) and *Ozette Archaeological Project Research Reports, vol. II. Fauna* (Pullman, WA: Washington State University and National Park Service, Pacific Northwest Regional Office, 1994), both edited by Stephan R. Samuels.

Further Visiting

Artifacts from Ozette are on display at the Makah Cultural and Research Center, located about 15 miles (24 kilometers) from Ozette in Neah Bay, Washington. Owned and operated by the Makah Indian Nation, the Makah Museum is the sole repository for archaeological materials discovered at the important coastal village of Ozette.

CHAPTER

SEVEN

———

Hopewell

100 B.C.—A.D. 400

Hopewell culture in Ohio

The origins of the Moundbuilders were the subject of a pitched debate until the nineteenth-century myth of a mysteriously lost race was laid to rest by a series of carefully planned, concerted archaeological excavations in the 1890s. One of these digs took place at Mordecai Cloud Hopewell's farm west of Chillicothe, Ohio. This astonishing site contained at least forty ancient conical mounds, enclosed by an extensive geometric embankment, obviously laid out with great precision.

As investigators began to explore the strange constructions, it became immediately clear that the ancient people who had built these mounds cared deeply about their ancestors. Not only had they built huge earthen monuments to encase their dead, but their respect was reflected in the magnificent objects placed in the graves: mica from the distant Appalachian Mountains, volcanic glass from Yellowstone, chalcedony from Knife River in North Dakota, conch shells and sharks' teeth from the south Atlantic, and copper from the Great Lakes. The Moundbuilders fashioned mystical and exotic artwork from these raw materials, then placed these offerings inside the tombs with their dead.

Artifacts recovered from Hopewell's farm were showcased at the Chicago's World Columbian Exposition of 1893. Since that time the site has been plowed into oblivion. The Hopewell site is today owned

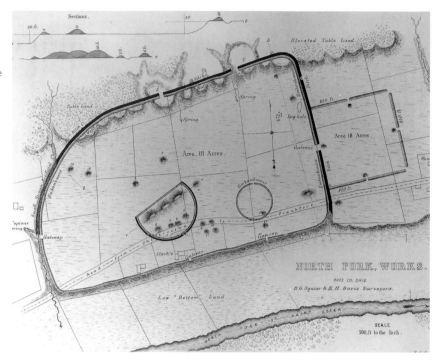

Captain Hopewell's Farm, called "North Fork, Works" as mapped by E. G. Squier and E. H. Davis in their classic book *Ancient Monuments of the Mississippi Valley*, which first appeared in 1848.

primarily by the Archaeological Conservancy, a national nonprofit organization formed in 1980 to identify, acquire, and permanently preserve the most significant archaeological sites in the United States.

The Mounds of Mound City

Not far from the Hopewell farm is Mound City, another major concentration of Hopewell burial mounds. It was not a city at all—it was a mortuary site for Hopewell people living along the Scioto River between 100 B.C. and A.D. 100. Here the modern visitor can get a sense of the dramatic funerals and lasting monuments Hopewell people erected for their deceased elite.

Mica Grave Mound, a spectacular trove of artifacts, was excavated in 1921. It yielded elk and bear teeth, copper ornaments, large obsidian projectile points, and a cache of five thousand shell beads. The mound takes its name from the thousand large, reflective mica disks found in it. Excavators also recovered two copper headdresses, one with three pairs of copper antlers, the other representing a bear with hinged ears and legs riveted on. The mound contained the remains of four individuals. As was the custom at Mound City, they had been cremated. Amid the ashes archaeologists found obsidian tools, effigy pipes depicting a raven and a toad, and a copper headpiece fashioned into a human form. For years, the contents of Mica Grave Mound were exhibited to show the elaborate multiple burial, but this exhibition has been closed permanently in response to Native American concerns.

This Hopewell platform pipe was recovered from Mound City, Ohio. The bird effigy is hollowed out to form the tobacco pipe bowl. The mouthpiece, with a thin drilled hole, runs through the base.

When the Mound of the Pipes was first excavated in the mid-nineteenth century, investigators recovered more than two hundred effigy smoking pipes depicting various animals, birds, and reptiles. Some think that the Mound of the Pipes was a monument to a master carver of sacred pipes, a fitting memorial indeed. Hopewell artists are famous for making pipes such as these from locally available pipestone. Best known are the platform pipes, which have cylindrical bowls resting upon straight or curved bases. Sometimes the bowls depict birds and animals.

Death Mask Mound, the largest and oldest of the two dozen mounds in this compound, held the remains of thirteen people, some of whom were accompanied by copper falcon effigies. This mound contained an original subterranean charnel house, which was later replaced by one built on the surface. Such charnel houses were often used to store the body of the deceased person. Sometimes a mound was erected atop the charnel house of a particularly important individual.

Mound City demonstrates how the system of social ranking worked in Hopewell society. The mounds themselves became visible reminders of the power of particular individuals and their families. The long period of planning, engineering, and execution involved in erecting such effective and enduring monuments was the group's investment in maintaining its own social order.

The rich grave goods of Mound City, presumably emblems of rank or status, also reflect Hopewell sociopolitical organization. Some archaeologists

Modern view across the central Hopewell mound group at Mound City, Ohio. Several reconstructed burial mounds and the encircling enclosure wall are clearly evident.

believe that animal effigies, such as those from Mound of the Pipes, represent long-vanished clans or lineages among the Hopewell people, similar to clans named after ravens and other animals by some later Native American groups (see Chapter 16 for a discussion of the northern Iroquoian lineage system). According to this interpretation, Hopewell society was divided into a series of rank-ordered lineages. If so, then each of the primary burial centers, such as Mound City, may have been used by one or more of the lineages as the final resting place for its most exalted individuals.

While Mound City is one of the most spectacular sites left to us by the Hopewell people, it represents only the mortuary aspect of the Hopewell lifeway. Mound City cannot be taken as "typical" of Hopewell life any more than Arlington National Cemetery is "typical" of late-twentieth-century life in the United States. At other archaeological sites, such as the Seip Mound in nearby Bainbridge, Ohio, one gets a different glimpse of the religious and the civic aspects of Hopewell society.

Two huge mounds once stood at Seip, surrounded by a large geometric earthwork west of Chillicothe; five smaller mounds stood nearby. Scientific excavation has revealed a number of Hopewell civic and ceremonial structures. When the buildings fell out of use, the people built a separate mound over each section; one of them is the second largest of the Ohio Hopewell mounds.

The Hopewell people also built earthen hilltop enclosures. The largest of these is found at Fort Ancient, southeast of Lebanon, Ohio. Here, on a bluff overlooking the Little Miami River, the Hopewell piled up more than 3.5 miles (6 kilometers) of earthen walls, enclosing more than 100 acres (40 hectares). Recent excavations have found evidence of a Hopewell village

WHAT DOES "WOODLAND" MEAN?

The umbrella term "Woodland" is confusing. Like "Archaic" and "Paleoindian," Woodland has a special meaning in American archaeology. The Woodland tradition is conventionally defined by the presence of three key traits or hallmarks: the manufacture of distinctive ceramics, the incipient development of agriculture, and the construction of funerary mounds. Throughout much of eastern North America—including the Midwest, the Southeast, the Northeast, and the eastern Great Plains—the Woodland period follows on the heels of the Archaic tradition.

Although its age varies considerably by location, most Woodland characteristics appeared by 1000 B.C. (or somewhat earlier) and lasted until replaced by other cultures, such as the Mississippian in the Southeast and the Plains Village Tradition on the Great Plains, between A.D. 700 and 1200. The Algonquian and Iroquoian traditions of the Northeast (ca. A.D. 1000–1650) are considered to represent a Terminal Woodland occupation. Elsewhere, as in the Far West, where so-called Archaic lifeways continued into the historic period, the Woodland characteristics never appeared at all.

The archaeological term "Woodland" is different from another, quite useful term: "Eastern Woodlands," which generally refers to the eastern half of Native North America, those millions of acres of primeval forests cut by countless coursing "river roads."

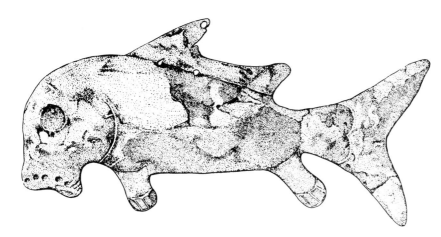

This naturalistic fish, representing a redhorse sucker or buffalofish, may have been regarded by the Hopewell people as a symbol of the underworld.

within the enclosure, along with evidence that the Hopewell builders filled in ravines and enlarged and leveled the hilltop before constructing the enclosure. Archaeologist Robert Connolly has estimated that the amount of earth moved to level and enlarge the hilltop may have equaled the amount of earth moved to build the walls of the enclosure: nearly 600,000 cubic yards (460,000 cubic feet).

The Hopewell Interaction Sphere

Burial mounds and plant domestication appeared in Mexico earlier than in what is now the United States. For this reason archaeologists long believed that mound building and agriculture must have arrived in the Southeast as a package from elsewhere. For years most archaeologists talked of Hopewell economics strictly in terms of growing maize. But better recovery and dating techniques now indicate that the Hopewell people produced a great diversity of locally domesticated foods long before Mother Corn arrived. There is also no direct evidence that any group ever moved from Mexico into the Southeast. Maize and beans arrived in the East at different times, apparently from the Southwest. Direct introduction from Mexico of either item is unlikely.

These new insights are important because they demonstrate that imported Mexican corn—perhaps by way of Basketmaker farmers in the American Southwest—caused neither the development of Hopewellian society nor a rapid shift toward agriculture in eastern North America. For centuries, maize was only a minor, almost invisible addition to already well-established food-producing economies. The extraordinary Hopewellian accomplishments in agricultural, religious, and belief systems resulted from homegrown ingenuity and inventiveness.

But how, precisely, did such a remarkable agricultural system evolve? Utilization of any particular plant crop was extremely localized, depending on local ecological circumstances. Foragers who relied very little on cultivated plants persisted in some places. But generally the broad midlatitude riverine

The most famous Hopewell artifact is the platform pipe, found in abundance at the Mound of the Pipes at Mound City. Some think that this mound was a monument to a master carver of sacred pipes.

Robert L. Hall has argued that archaeologists have spent so much time emphasizing economy and ecology that they entirely overlook the symbolic and "affective" qualities of these unforgettable Hopewellian artifacts. What do they have to tell us? Why, for instance, is the "peace pipe" used historically to establish friendly contact almost always in the form of a weapon? Why did the famous Hopewell pipes take the forms they did?

This Hopewell stone effigy pipe represents a beaver. It was found in the Bedford Mound in Illinois.

To explore such questions, Hall began with a model familiar to all. He reasoned that everyone engages in certain culturally dictated customs whose exact meaning and origin may be lost in time. For example, the rite of "toasting" originally was the sloshing and spilling together of two persons' drinks to reduce the possibility that one planned to poison the other. But how many of us who have toasted friends realize the origins of the custom? Similarly, the salute stems from the act of raising visors on armored helmets in order to expose the faces of the two persons encountering each other. Although these gestures lost their original, practical significance, the acts survived as elements of etiquette or protocol.

Hall decided to apply similar reasoning to the Hopewell platform pipes. Throughout historic times in the eastern United States, Indian tribes observed the custom of smoking a sacred tribal pipe. When the pipe was present, violence was absolutely ruled out. Moreover, the so-called peace pipe (the calumet) usually was made in a distinctive, weaponlike form. The Pawnee peace pipe, for instance, looked like an arrow. In fact, the Osage word for calumet translates as "arrowshaft." Hall suggested that the weaponlike appearance resulted from a specific ceremonial custom. Could it be that, at least during the period of European contact, the peace pipe symbolized a ritual weapon?

Hall then extended his hunch back into ancient Hopewell times. Suppose that the distinctive Hopewell platform pipes were also ritual weapons, made before these people knew of the bow and arrow. At that time, the most common Hopewell weapon was the atlatl, or spear thrower. Hall suggested that the distinctive Hopewell platform pipe symbolically represented a flat atlatl, decorated with an effigy spur. He observed that the animal on the bowl was almost always carved precisely where an atlatl spur would be, and the curvature of the platform seemed to correspond to the curvature on the atlatl.

zone—stretching from the Appalachian wall west to the prairie margin—was the homeland of early food-producing societies.

These Native people grew squash, marsh elder, sunflower, goosefoot, erect knotweed, and maygrass. Modern experiments have demonstrated the previously unappreciated economic potential of these indigenous eastern North American crop plants (which are now available in most modern health food stores); harvest yields are fully comparable to the various European wheats cultivated during the nineteenth century.

From about A.D. 1 to 200, Hopewellian habitation sites remained small—

These correspondences led Hall to conclude, in a 1977 article in *American Antiquity*, "I see the Hopewell platform pipe as the archaeologically visible part of a transformed ritual atlatl, a symbolic weapon which in Middle Woodland times probably had some of the same functions as the calumet of historic times, itself a ritual arrow." He went on to propose that the importance of the Hopewell pipe might well extend beyond mere symbolism. The platform pipe was not merely one of many items exchanged between groups, but "may have been part of the very mechanism of exchange." And here is the potential contribution of Hall's work.

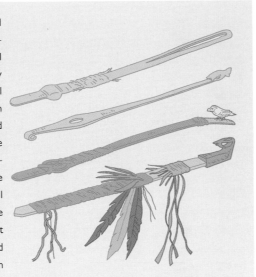

Several kinds of atlatls (spear throwers) used in Native North America

Research on eastern United States prehistory has conventionally defined the Hopewellian Interaction Sphere primarily in economic and environmental terms. Perhaps, by maintaining relationships between large-scale networks of ritual trading partners, far-flung Hopewell communities could have joined economic forces, looking to one another for support in lean years of drought or crop failure. Hall suggested that a shifting away from strictly materialistic thinking—toward a new, cognitive approach—could generate a broader understanding of the Hopewell lifeway. Reasoning from historically recorded Native American analogies, Hall contended that peace pipe ceremonialism served to mediate interaction over a vast central part of the United States and Canada. Hall was not suggesting that researchers ignore the economic and political ramifications of such interaction, but he did urge archaeologists to also consider the symbolic details of Hopewellian exchange. He argued that through "peace pipe diplomacy," the Hopewellian Interaction Sphere tended to reduce regional differences and promote friendly contact and communication between discrete groups.

generally one to three household settlements dispersed along stream and river valley corridors. In some floodplain lakes and marshes, such as the lower Illinois River, small household settlements formed loose spatial concentrations that can appropriately be termed "villages," even though they lack any suggestion of an overall community plan.

The broad-spectrum and flexible premaize economies of these household units also demonstrate the degree to which production of storable food harvests was simply overlaid on an already successful foraging pattern. The domesticates simply provided a hedge against potential food shortages. There was, in short, no wholesale replacement of the earlier hunting-and-gathering ecology.

The Hopewell complex collapsed by A.D. 400, for reasons not entirely understood. Perhaps this breakdown was due to a collapse in traditional trading relationships. Or possibly the Hopewell demise came about from warfare and social unrest. Whatever the cause, it is clear that the Hopewell people did not disappear or necessarily migrate elsewhere. One way or another, they were ancestral to the fully agricultural, hierarchially structured society that archaeologists call Mississippian.

Regionalized Interactions: The Great Hopewell Road?

The physical distribution of distinctive Hopewell artifacts came to reflect the realm of Hopewell ideology and religion. Yet despite widespread commonality, there was no single person or individual group who ruled the entire Hopewellian territory. Instead, each individual region remained under local control. Archaeologists seem to know more about long-distance Hopewellian exchange than about ritual and interaction at the local level. How did the individual Hopewell polities interact?

One possibility is suggested by the practices of Native American tribes during the historic period. Tribes often gathered together for feasts, ceremonies, and contests of skill, while individuals conducted their own private business on the periphery. We know that the Hopewell erected several different kinds of earthworks: village, burial, and some apparently used for other purposes, probably religious ceremonies and games or political transactions. Can we see this regional ritual interaction in these nonmortuary earthworks?

Recent research being conducted at the Newark earthworks suggests an intriguing possibility. These earthworks, on the west side of Newark, Ohio, were originally among the most extensive in North America—certainly the largest of the Hopewell world. Today, the city of Newark has sprawled over large parts of the Hopewell earthworks, obliterating walls and mounds. But important sections are still preserved and administered by the Ohio Historical Society.

The geometric constructions at Newark and elsewhere in Ohio must have formed an important part of Hopewell religious and social life, probably also for periodic rites of renewal. The earthworks may have been used for formal meetings of state, religious ceremonies, and perhaps also as venues for individuals to meet face to face and conduct private transactions. The Newark earthworks were assembled on a grand scale, presumably reflecting the importance of the activities held within their confines. The original earthworks covered 4 square miles (10 square kilometers), including several large compounds connected by causeways.

Bradley Lepper, archaeologist and curator of the Newark earthworks, has come up with a startling new theory. Well aware of the newly discovered road

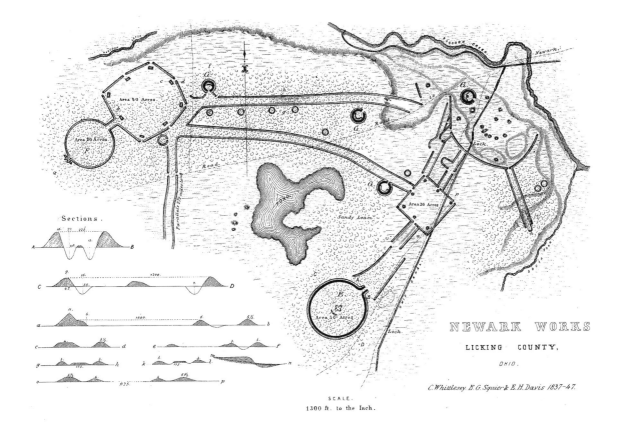

Sections.

NEWARK WORKS

LICKING COUNTY,

OHIO.

C. Whittlesey. E. G. Squier & E. H. Davis 1837-47.

SCALE.
1300 ft. to the Inch.

system connecting Chaco Canyon to its outliers (see Chapter 11), Lepper began looking for a similar system of roads that might once have connected the important ceremonial centers of the Hopewellian heartland. He may have found it. Digging through various nineteenth-century records, Lepper turned up four different accounts documenting a 6-mile (10-kilometer) stretch of road that may once have connected the major Hopewell monuments at Newark with those near Chillicothe 60 miles (100 kilometers) away. Caleb Atwater was the first to mention the road in 1820, describing a set of parallel earthen walls projecting up to 30 miles (50 kilometers) southwest from the major monuments at Newark. A 1862 account traces more than 6 miles (10 kilometers) of roadway, apparently extending toward Chillicothe. Unfortunately, cultivation and construction have today obliterated most traces of the ancient roadway, if it existed at all.

A rare aerial photograph from the 1930s seems to show two clear lines cutting an arrow-straight path across a patch of Licking County farmland. As he flew over the suspected route, Lepper was electrified to see below two parallel linear shadows in precisely the same location as on the 1934 photograph. When he explored the area on the ground, he found remnants of standing walls—the first tangible evidence of the Great Hopewell Road.

Map showing the projected course of the Great Hopewell Road, running between Newark and Chillicothe in southeastern Ohio. The four numbered areas indicate where possible traces of the road have been observed.

Lepper projected the road course along a 31-degree west of south bearing from Newark to Chillicothe and used additional aerial reconnaissance and archival photographs to search the corridor for further traces. He found several traces of parallel linear features, especially visible as infrared images. He reconstructed the Great Hopewell Road as consisting of parallel walls of earth 3 feet (1 meter) high and nearly 200 feet (60 meters) apart.

Lepper contends that this roadway connected two important destinations in the Hopewell world. Each site contained a circular embankment connected to an octagonal enclosure; they are built nowhere else. Although built more than 55 miles (90 kilometers) apart, they are oriented at exactly 90 degrees to one another. Both encode considerable astronomical information, including a knowledge of the 18.6-year lunar cycle. So by building the great road, the Ohio Hopewell physically connected two sites that were already intellectually connected in many ways.

How would such a ritual roadway work? Perhaps, as when leaders of the modern world meet, the ancient Hopewell exchanged presents, particularly gifts representative of the crafts and handiwork of their home countries. Such visits establish new contacts between nations and also between different sectors of the same nation, such as bureaucrats, businessmen, religious leaders, and private citizens. What begins as symbolic exchange between rulers can trickle down to become formal patterns of commerce. The same may have been true in Hopewell times. Some archaeologists think that certain of the amazing artifacts recovered from places like Mound City may have been

This 1930s aerial photograph shows the Newark Earthworks (at the left of the photograph), and two telltale traces of the Great Hopewell Road (highlighted by the two circles).

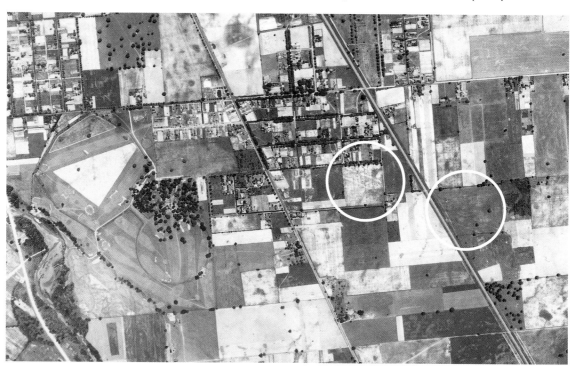

exchanged to establish commercial and diplomatic ties between two distant political groups. Perhaps the ritual road at Newark was once used for processions carrying such diplomatic items. Of course, Lepper's findings must remain tentative until more concrete archaeological evidence is available, and some archaeologists are suspending judgment. Still, the distinct possibility of the Great Hopewell Road remains exciting.

Further Reading

Several useful summaries of Hopewell archaeology are available, including "Tracking Ohio's Great Hopewell Road," *Archaeology* 48, no. 6 (1995): 52–56, and *People of the Mounds: Ohio's Hopewell Culture* (Chillicothe: OH: Hopewell Culture National Historical Park, 1995), both by Bradley T. Lepper; *Hopewellian Archaeology*, edited by David Brose and Greber, N'omi (Kent, OH: Kent State University Press, 1979); *The Hopewell Site* by N'omi Greber and Katherine C. Ruhl (Boulder, CO: Westview Press, 1989); and *A View from the Core: A Synthesis of Ohio Hopewell Archaeology*, edited by P. Pacheco (Columbus: Ohio Archaeological Council, 1996).

More general context for the Hopewell complex can be found in *Rivers of Change: Essays on Early Agriculture in Eastern North America* by Bruce D. Smith (Washington, DC: Smithsonian Institution Press, 1992); *Ancient Earthen Enclosures of the Eastern Woodlands* by Robert Mainfort and Lynne Goldstein (Gainesville: University of Florida Press: 1998); and *The Moundbuilders* by Robert Silverberg (Greenwich: New York Graphic Society, 1968).

Further Visiting

The Hopewell Culture National Historical Park (Chillicothe, OH; about 3 miles [5 kilometers] north on SR 104) was established in 1923 to protect a group of Hopewell mounds in a 13-acre (5 hectare) square enclosure. A visitor center includes exhibits and an observation deck. Newark earthworks (Newark, OH) contains preserved portions of this massive system of geometrical earthworks. Moundbuilders State Memorial contains a circular earthwork 1,200 feet (365 meters) in diameter, with earthen walls 8 to 14 feet (2.5 to 4 meters) high. The Ohio Indian Art Museum exhibits ancient cultures of Ohio in various media.

The so-called Wray figurine probably represents a Hopewell shaman. It was recovered from one of the principal burial mounds at Newark Earthworks.

93
—

Poverty Point

1730–1350 B.C.

Poverty Point culture

in Louisiana

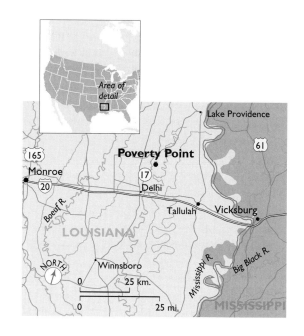

The Poverty Point site has long been a headache for American archaeologists. The earliest investigators simply walked over the most massive and widespread earthworks, mistaking them for natural levee ridges. Only after aerial photographs became available did archaeologists realize that people had built these huge, concentric earthworks. So large were these constructions that, as historian Roger Kennedy once remarked, seven Stonehenges could be erected inside.

But once the true extent of Poverty Point was recognized, the real problems began for archaeologists attempting to understand what had taken place here. Not only were the huge earthworks really old—the radiocarbon dates cluster around 1500 B.C.—but investigators failed to find any evidence for agriculture at Poverty Point. This was a big problem because archaeologists have long believed that non-agricultural hunting and gathering people lived in "unsophisticated" small bands or rudimentary tribes and were wholly incapable of joining together for large-scale community projects. According to this conventional wisdom, monumental construction efforts such as those at Poverty Point can be undertaken only by relatively large populations, living in permanent villages and supported by a fully agricultural economic base.

So how does Poverty Point fit with the rest of North American archaeology? To some investigators, the absence of agriculture means Poverty Point must belong to the Archaic stage (using the criteria for Archaic set out in Chapter 2). But other archaeologists looked at the massive earthworks and believed that Poverty Point must be classified as Formative, more like the Hopewell and later cultures of the eastern Woodlands (using the criteria for Woodland set out in Chapter 7). Perhaps the evidence for farming and social stratification had simply been overlooked at Poverty Point.

Poverty Point stuck in the craw of American archaeology for decades. It was either a taxonomic oddity or a precocious harbinger of things to come. It stood alone, an archaeological enigma.

Early Explorations

The first "archaeological" exploration at Poverty Point seems to have taken place in the twelfth century. Modern archaeologists excavating in northwestern Louisiana discovered a twelfth-century Caddo Indian buried with his medicine bundle, a skin bag containing two tubular red stone beads, a slate pendant, and a perforated plummet of polished hematite. Without question, these ancient heirlooms came from the Poverty Point site, or perhaps some other long-abandoned site built by the Poverty Point villagers.

The Poverty Point site was first formally recorded in 1873 by Samuel Lockett, who noted the presence of several large earthen mounds built along the eastern front of Macon (pronounced "Mason") Ridge, a 25-foot (8-meter) high bluff that defines the edge of the Mississippi River floodplain in northeastern Louisiana.

Serious archaeological exploration did not begin until 1912, when Clarence B. Moore piloted his riverboat *Gopher* up Bayou Macon, stopping at the Poverty Point site to prepare a detailed map of what he termed Mound A. Moore was perplexed by the apparent absence of pottery, but he did collect and beautifully illustrate the curious artifacts that characterize this unusual site: baked clay balls ("Poverty Point objects"), hematite plummets, a clay figure, jasper beads, steatite sherds, and some large projectile points. Gerard Fowke followed up with a Smithsonian-sponsored exploration in 1926, and sporadic investigations continued through the 1930s and 1940s.

Clarence H. Webb began his long-term association with the site in the 1935. A decade later, James A. Ford headed a team that carried out three seasons of major excavations at Poverty Point. They mapped the mounds, confirming the accuracy of Moore's previous measurements. They also obtained several radiocarbon dates suggesting that the earthworks were erected quite early, between 1200 and 400 B.C.

The Louisiana State Parks Commission acquired the site in the early 1970s, and further excavations were undertaken when it was converted to a state commemorative area. Numerous field schools from the University of

This oblique aerial view of the Poverty Point site was taken by Junius Bird in the mid-1950s (looking north-west). The Mississippi River floodplain and the Bayou Macon are evident at the lower right, and the massive concentric rings stand out dramatically.

Southwestern Louisiana and elsewhere also helped to work out the cultural chronology and building sequence described here.

The Mounds

The largest mound at Poverty Point measures 710 by 640 feet (215 by 195 meters), and stands 70 feet (20 meters) high. Fowke mistakenly believed this mound was a natural formation, a geological remnant, but Ford's systematic excavations clearly established that it had been built by human hands— 185,000 cubic yards (140,000 cubic meters) of fill laboriously piled up, one basketload at a time. Ford estimated that building such a mound would require more than three million man-hours of labor. The mound stands as tall as a seven-story building, the second highest Indian earthwork in North America after Monks Mound at Cahokia (see Chapter 13).

To Ford, the main mound at Poverty Point resembles a huge bird, sitting with wings outspread. Elongate ovals form the body and wings, the upthrust mound crest is the head, and the extended platforms simulate the tail. He marveled at the geometric accuracy and the solar orientation of the major mounds. The projecting ramp of the major mound is within 7 degrees of due east, and that of the Motley Mound is similarly pointed nearly due south.

Ford and Webb also extensively explored Mound B, a smaller conical construction some 700 yards (640 meters) north of the Poverty Point Mound. They found that it had been erected in three clearly defined stages. At the end of each stage, the top was leveled, and some ceremony was presumably held there. After completing the second level, the builders discarded several hundred baskets and skins filled with earth. Casts of these burden baskets were

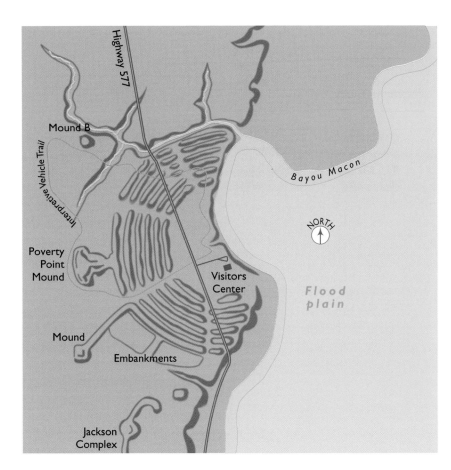

The distribution of earthworks at Poverty Point. The Jackson site, along the southern margin, postdates the late Archaic geometrical earthworks.

so well preserved that Ford's team was able to reconstruct the weave, shape, and size of the baskets used to haul the 860,000 to 980,000 cubic yards (660,000 to 750,000 cubic meters) of earth contained in the Poverty Point earthworks.

Ford and Webb exposed a premound habitation surface beneath Mound B, covered by a thick ash bed that contained charred bone. They thought this

Stratigraphic profile of the north wall of Mound B at Poverty Point. The black layer of Floor I is an ash bed 6 inches (15 centimeters) thick, containing charred human bone.

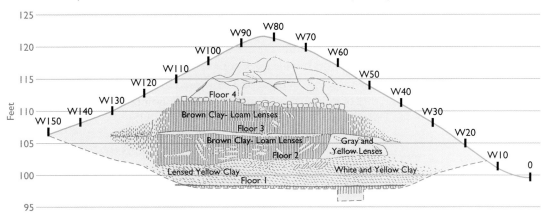

Excavating the 70-foot (20-meter)-long trench through Mound B at Poverty Point. The excavators are standing on the hard-packed ash bed (Floor 1).

ash bed and burnt human bone must be the remains of an important personage who had been cremated there, with the mound erected over the charred charnel house. The radiocarbon dates for the ash bed suggest that the great fire had been kindled about 1730 to 1350 B.C., with the mound erected sometime thereafter. This date was critical because it suggested that Poverty Point might have been partly contemporaneous with the better-known Hopewell cultures of the Midwest.

The Concentric Rings

Even untrained eyes find it is difficult to miss the major mounds at Poverty Point. But the massive set of low-lying geometrical ridges is not so easily recognized. It was not until Ford examined aerial photographs of the site in 1953 that archaeologists recognized the huge C-shaped earthenworks, three-quarters of a mile across. This elliptical ridged enclosure comprises six concentric raised earthen rings, spaced some 150 to 185 feet (45 to 55 meters) apart crest to crest. Today they stand 5 to 10 feet (1 to 3 meters) high. Each ridge is divided into six compartments by five transecting aisles. In addition to these above-ground earthworks, large quantities of fill leveled the uneven construction area, creating the distinctive swales or ditches between the ridges.

Ford and Webb thought that the Poverty Point earthworks had originally been constructed in a symmetrical, octagonal configuration, and that the eastern portion had subsequently been eaten away by the Bayou Macon. But recent geological investigations have shown that the bluff along the eastern margin of the site had been there for thousands of years before construction began. This means that the concentric rings must have been deliberately designed as a semicircular form from the beginning, built according to a detailed master plan.

Since the Ford and Webb excavations, most of the mounds have been tested and redated. More than three dozen radiocarbon dates are available for the Poverty Point earthworks, ranging from about 4000 B.C. to A.D. 960. But most dates cluster around 1730 to 1350 B.C., providing the currently accepted estimate for the major on-site activities, including earthwork construction.

Artifacts from Poverty Point

The most abundant artifact is the so-called Poverty Point object, hallmark of the entire complex. Ford once calculated that perhaps 20 million such objects were made and discarded by the inhabitants of Poverty Point. Often called "clay balls," Poverty Point objects are neither balls nor clay—they're actually made of silt. These objects were generally handmolded, often by children. Makers used fingers, palms, and occasionally tools to fashion dozens of individual styles: cylindrical, melon-shaped, biconical, pyramidal, biscuit-shaped, pillow-shaped, even mushroom-shaped.

So what are they? Ford and Webb concluded that the "clay balls" were artificially constructed cooking stones. The Poverty Point site was situated in an alluvial valley some 30 miles (48 kilometers) from the nearest source of stone. In this stone-poor environment, the Poverty Point objects were apparently intended for use in earth ovens. Ford and Webb noted that the Poverty Point objects tended to cluster around hearths, and some were found still packed inside undisturbed firepits. Apparently the people created an oven by digging a hole, packing the "clay balls" around the food, and then covering the pit. Such ovens both regulated heat and conserved energy. It was thought that the sizes and shapes of the Poverty Point objects controlled how hot the pit got and how much heat was retained, and some investigators believe that cooks fine-tuned the cooking temperature by using objects of different shapes. But a recent engineering study indicates that neither size nor shape affects heat storage and transfer capacities.

The octagonal earthworks at Poverty Point were also littered with whole and broken spear points; 25 to 30 percent of these were made of local pebble chert, the rest from imported Arkansas or midwestern flint. Thousands have been collected.

Other diverse and distinctive artifacts belonging to the Poverty Point complex include tubular pipes, baked clay human figurines, stone vessels,

Various "Poverty Point objects" (baked clay cooking balls). The Poverty Point site may have originally contained 24 million such objects.

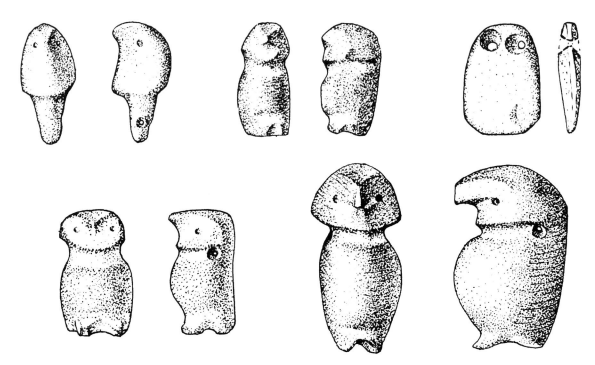

Tiny owl effigies made of jasper from Poverty Point. The largest is .75 inches (2 centimeters) high. Most have small drilled holes for stringing.

microlithic tools, rough greenstone pseudo-celts, plummets made of iron minerals, and polished stone lapidary beads (particularly jasper). This artifact assemblage not only demonstrates a high degree of craftsmanship but also establishes a distinctive pattern of long-distance trade: copper from the Great Lakes; lead ore (galena) from Missouri; soapstone (steatite) from Alabama and Georgia; and various toolstones (used for dart points and knives) from the Ouachita Mountains of Arkansas, the Tennessee River Valley, the Shawnee Hills in southern Illinois, the Ozark Plateau in eastern Missouri, the Knobs in western Kentucky, and the Appalachian Piedmont of Alabama and Georgia.

The Poverty Point Problem

According to Ford and Webb, Poverty Point was first occupied by a resident Archaic population that exploited the rich neighboring swamplands and perhaps numbered in the thousands. An invading group of Hopewell farmers from the uppermost drainages of the Mississippi Basin overwhelmed this pioneering hunting and gathering population. The Hopewellian invaders operated on strictly organized theocratic principles and brought with them an agricultural economy. The surviving Archaic population at Poverty Point was reduced to lower-class laborers and forced to construct the massive earthworks, perhaps to honor the faraway Hopewell gods.

Other researchers saw the same evidence differently. For Gordon Willey, Poverty Point was not an integrated cultural complex at all. It was instead a palimpsest, an archaeological hodge-podge left by many unrelated moundbuilding populations. Perhaps the site was actually just a temporary encampment, a vacant ceremonial center similar to the geometric Hopewell earthworks and sacred circles. Willey would later abandon the vacant center hypothesis, suggesting instead that Poverty Point was home to a large, sedentary, maize-growing population with roots ultimately extending into Mesoamerica.

Ford and Webb revised their position in the late 1960s, suggesting that Poverty Point arose when some Archaic groups of the lower Mississippi Valley came into contact with Formative people from Vera Cruz, resulting in a burst of cultural energy grounded in maize agriculture and theocratic social stratification. Hopewell developments in the upper Mississippi Valley were seen as ultimately deriving from Poverty Point, rather than the other way around.

Each of these interpretations reflected the long-standing belief that mound building, pottery, agriculture, sedentism, and large populations were passed along as an integrated complex. In this view, a single developmental sequence could be applied across all of Eastern North America: Paleoindian and Archaic foragers did not build earthworks of any kind. Burial mounds appeared in Woodland times (probably initiated by midwestern Hopewell farmers) and eventually evolved into the flat-topped, rectangular temple mounds of the Mississippian period.

This is why Poverty Point posed such a problem for the pan-Eastern sequence. Although the huge earthworks resembled those built by the Hopewell, the artifacts were clearly Archaic. The Hopewellians were farmers, the moundbuilders at Poverty Point were apparently not. Poverty Point remained an archaeological mystery for decades.

Poverty Point: First North American Chiefdom?

So things stood in 1970 when Webb invited Jon Gibson to help analyze a huge collection from the site. More than 18,000 artifacts were classified into typological, functional, and material source categories, then regrouped into analytical categories based on the inferred gender and social class of the user.

Several distributional patterns emerged, leading Gibson to suggest that the Poverty Point site contained all the key diagnostics of chiefdom-level social organization: a large, sedentary, hierarchically organized population with craft specialization, governed by a small chiefly elite. Gibson saw the Poverty Point site as a socioeconomic and ceremonial center surrounded by dozens of lower-ranked affiliates. The highly productive hunting and gathering economy produced a surplus, first funnelled to an elite, then redistributed to commoners. So viewed, Poverty Point was the earliest mainland chiefdom in North America.

Although Gibson's work was a major step forward, some cracks eventually appeared in his reconstruction. The chiefdom hypothesis was grounded in the sociopolitical model developed by Elman Service, Marshall Sahlins, and

EARLY SOUTHEASTERN MOUND BUILDERS

James Ford and Clarence Webb established Poverty Point as the earliest mound complex in North America, centuries older than the widespread Adena and Hopewell mound complexes of the American heartland. Throughout the 1960s, archaeologists working in other mound sites in Louisiana came up with curiously early Archaic-age dates. But the data was always just suggestive, and the antiquity of these sites remained in doubt. More such sites were dated during the 1970s, and results suggested that Louisiana and Florida might have earthworks older than five thousand years. Finally, by the 1990s, four additional mound sites in Louisiana and two more in Florida were securely dated to the Middle Archaic period.

The Watson Brake site in northeastern Louisiana is the largest, most complex, and most securely dated. Watson Brake contains eleven mounds connected by ridges to form an oval-shaped earthen enclosure 920 feet (280 meters) in diameter. Gentry Mound, the largest, is 25 feet (7.5 meters) high. Multiple radiocarbon dates indicate that Watson Brake was built from 5400 to 5000 B.P.

For decades, planned large-scale earthworks such as Watson Brake were considered to be far beyond the leadership and organizational abilities of seasonally mobile, nonagricultural people. Poverty Point was regarded as an exception, and its extensive trade network was evidence for sophisticated socioeconomic organization. But now the Watson Brake earthworks are known to predate those at Poverty Point by nearly two thousand years, and combined floral and faunal data leave little doubt that the Watson Brake constructions were built and occupied by hunter-gatherers. These early moundbuilders fished from early summer through the fall, also collecting seeds of weedy plants, especially goosefoot and knotweed. They made fired earthen objects (so-called Poverty Point objects) in a variety of shapes; several are so highly fired that they can legitimately be called "ceramic." They used only locally available gravel cherts to make stone tools.

Morton Fried. Drawing upon the available ethnographic evidence, these authors assumed that the chiefdom-defining characteristics should appear as a unified complex. The chiefdom model began to unravel when archaeologists started turning up evidence elsewhere throughout North America that sedentism and cultural complexity appeared in nonagricultural contexts as well. The so-called hallmarks of cultural complexity—high population density, permanent ceremonial ground and cemeteries, differences in grave goods graded according to wealth, construction of burial mounds and earthworks—were showing up in the apparent absence of an agricultural economy. In other words, hunter-gatherer economies could no longer be regarded as monolithic and self-limiting. This meant that the sociocultural complexity of Poverty Point could not be reconstructed by strict reference to ethnographic archetypes.

Another Light on Poverty Point

Recent investigations at Watson Brake and elsewhere have demonstrated that nonagricultural peoples of the lower Mississippi Valley were raising significant earthworks more than five thousand years ago. Clearly, mound building was an old, established tradition before the first earthworks arose at Poverty Point. No longer can the Poverty Point earthworks be viewed as the earliest such constructions. They are merely the largest and most complex in a long continuum.

Gibson has proposed a new model to explain the complexity at Poverty Point. For one thing, it is now clear that the Poverty Point site was occupied before the major earthworks were erected. Like the people living at Watson Brake, the earliest Poverty Point residents made their stone tools strictly from local gravels. Their economy was grounded in hunting, collecting, and especially fishing in nearby waters.

Gibson argues that backwater fishing was so extraordinarily productive that the best fishermen could put away a considerable surplus, often feeding neighbors and kinfolk, who then found themselves in debt. Short-term debt could readily be paid off as economic fortunes turned. But over the longer run, some debtors were required to work off their obligations through community service such as constructing public earthworks. Building mounds could have served to foster community pride and ethnic identity, to memorialize tribal history, and to mark the boundaries of tribal land. At least one mound, the conical Lower Jackson Mound, was built by these early inhabitants.

Gibson believes that "special" Poverty Point-defining elements were already present during this earlier Archaic occupation: heavy emphasis on fishing, social inequality, and a self-perpetuating cycle of aggrandizement through mound construction. The lone missing component was long-distance exchange.

About 1700 B.C., the various occupation areas at Poverty Point consolidated into a large half-ring, 5,500 feet (1,700 meters) across and 2,000 feet (600 meters) wide. Open to the east, this broad C-shaped ring of midden terminated along a towering bluff overlooking the Mississippi swamplands. For whatever reason, the onset of long-distance exchange coincided with this settlement pattern shift. The entire range of exotic materials—hematite, magnetite, slate, copper, and so forth—is found in this large midden ring. Serious mound construction began shortly thereafter as residents erected the six concentric embankments, exactly covering this ring of occupation midden.

What triggered the long-distance trade? Gibson suggests that the impetus was a major shift in fishing technology. By tying stone weights to the bottom of their gill nets and seines, fishermen could work in a variety of water conditions, expanding the range of potential fishing areas from the slack backwaters and sluggish bayous to rapidly flowing streams, thus extending the potential fishing seasons each year. As this new technology created larger surpluses, debt obligations intensified and community-works projects flourished.

Stone plummets from
Poverty Point helped extend
the fishing season.

The best net sinkers, called plummets, were made of the heavy iron ore (hematite and magnetite) found 200 miles (320 kilometers) from Poverty Point, up the Arkansas River in the Ouachita Mountains near Hot Springs, Arkansas. The Poverty Point site was well-situated strategically, located on the first high land near the Arkansas and Mississippi rivers. This strategic location proved perfect for the long-distance trade networks that eventually developed.

But procuring iron ore over such long distances required considerable planning, probably by the same successful families who supported the initial mound construction—or, perhaps, by their new competitors. At Poverty Point iron ore and dozens of other exotics and imported materials were distributed to everyone who required them, not just the privileged few. How this distribution took place is unclear, but it would seem that gifting and obligation became the lifeblood of those living there. In a paper delivered to the annual meeting of the Society for American Archaeology in 1996, Gibson explained the process:

> With so much at stake, aggrandizers sought to protect their investment from disruptive supernatural forces that were everywhere. They called in markers on an enormous scale. They had a succession of protective earthen rings built around Poverty Point's inner sanctum and, as a backup, encased the six, concentric shields within two mound-defined axes that guarded the west and north sides, the directions from which witches, ghosts, and other evil forces operated. The cost was enormous, nearly seven million manhours of labor.

In this view Poverty Point was not a vacant ceremonial center. Nor was it conquered by Hopewellian battalions or missionaries from ancient Mesoamerica. Poverty Point was the largest town around, home to a corporate community, economic and spiritual hub of a much larger region.

Further Reading

Roger G. Kennedy provides a refreshing introduction to Poverty Point archaeology in *Hidden Cities: The Discovery and Loss of Ancient North American Civilization* (New York: Free Press, 1994) and a classic introduction by Jon L. Gibson appears in "Poverty Point, the First North American Chiefdom," *Archaeology* 27, no. 2 (1974): 96–105. Other important contributions include "Archaic Mounds in the Southeast," edited by Michael Russo, *Southeastern Archaeology* 13, no. 2 (1994): 89–162; *Poverty Point, A Late Archaic Site in Louisiana,* by James A. Ford and Clarence H. Webb (New York: American Museum of Natural History, 1956); and *Poverty Point: A Culture of the Lower Mississippi Valley* by Jon L. Gibson (Baton Rouge, LA: Louisiana Archaeological Survey and Antiquities Commission, Anthropological Study 7, 1983).

Further Viewing

Poverty Point State Commemorative Area (Epps, LA; 4¼ miles [7 kilometers] east on SR 134 and 1 mile [1.5 kilometers] north on SR 577) is a 400-acre (160-hectare) preserve, with a museum, observation tower, and interpretive walking and vehicle trails. Occasionally visitors can watch excavations in progress.

NINE

Serpent Mound

A.D. 1000 – 1140

Fort Ancient tradition

in Ohio

One of America's most intriguing archaeological artifacts does not fit inside a glass museum case. It is a mounded serpent nearly one-quarter of a mile long, mysteriously crawling along a bluff overlooking Brush Creek in southwestern Ohio. The ancient builders of the Great Serpent Mound carefully planned this oversized effigy by first outlining with small stones and lumps of clay a monstrous snake. Then they piled up countless baskets of yellow clay over the outline, burying their markers.

The result is a flawlessly modeled serpent, forever slithering westward toward the summer solstice sunset. Many observers think the snake may be trying to swallow an egg, judging from the oval earthwork near the mouth. Others suggest that the oval is a symbolic representation of the reptile's heart, or perhaps an enlarged and highlighted eye.

Some Cosmological Perspectives

The Serpent Mound is pure art. It amazes. It has power. But why was it built? And what does it mean?

The serpent has always been the most mysterious of all creatures. Alone among the animals, it is swift without feet, fins, or wings. The snake sheds its skin every spring and to many symbolizes the annual

renewal of life. For many Indian people, the cast-off snakeskin has the power to cure and heal.

While slow to attack, the rattlesnake is venomous in the extreme. It is like lightning: the quick spring, zigzag course, and rapid strike, the mortal bite. For the Hopi, the Great Horned Serpent or Lightning Snake has udders, conforming to tribal tradition that all of the earth's water and blood came from the breast of a single great serpent. This mythological serpent rules over the terrestrial waters, controlling the most important element for an agricultural people in a dry landscape. The snake also rules floods, earthquakes, and landslides. It can inflict punishment, particularly for sexual misconduct. Some Pueblo people dance with live rattlesnakes in late summertime. Ceremoniously handled, the snakes are released unharmed into the desert, so that they may convey to Mother Earth the prayers of those needing water.

Do these rituals and beliefs inform us about the curious Serpent Mound in Ohio?

The Ohio Indians of generations ago saw the snake as symbolic of Ursa Major and Ursa Minor, the Big and Little Dippers. Astronomers studying Serpent Mound point out that the serpent's body is congruent with the shape of the Little Dipper, that the movement apparent in the serpent's tail reflects the progress of the constellation around the North Star.

And what about the egg? Is there any relationship to the ancient Asian legend of a serpent swallowing the moon to create a lunar eclipse?

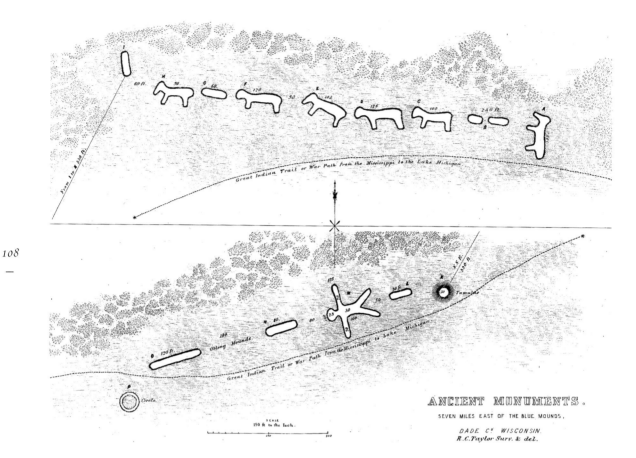

ANCIENT MONUMENTS.

SEVEN MILES EAST OF THE BLUE MOUNDS,

DADE Cᵒ WISCONSIN.
R.C.Taylor Surv. & del.

Squier and Davis drawings of effigy mound group, Dade County, Wisconsin

Early Archaeological Musings

It took the near-destruction of Serpent Mound to save it.

Even before an epic archaeological survey by Ephraim Squier and Edwin Davis in 1848, at least one major act of vandalism had already marred Serpent Mound, perhaps commited by an early gold-seeker. A decade later, a tornado struck the site, uprooting most of the old-growth trees growing on the serpentine earthworks. Once the tree-cover was destroyed, the area was soon cleared for cultivation and livestock grazing.

Frederick Ward Putnam, then of Harvard University, first visited Serpent Mound in 1883. When he returned three years later, he was shocked to see that repeated plowing was systematically demolishing the sinuous mound. Believing that Serpent Mound would soon be destroyed unless somebody intervened, Putnam wrote in a letter:

> It was with sorrow that we noticed the injury which had been done to the work since our last visit [1883]. Hundreds of persons visit the place every year, and among them have been vandals who have dug into the embankment and left unfilled the holes they made. As a con-

sequence the rains the tramplings of cattle and visitors, have caused such places to wear away and thus seriously injure this sacred work, the only one of its kind, and now universally known. It is sad to realize that this ancient work will probably be a thing of the past if another season is permitted to go by without the work of destruction being stayed.

Putnam was successful in raising money to purchase the Serpent Mound and an adjoining 75 acres (30 hectares) for Harvard University.

Putnam spent three field seasons (1887–1889) working at Serpent Mound and the surrounding sites. He excavated a series of trenches through the oval embankment and body area of the serpent, demonstrating the degree to which the original ground surface had been prepared for the subsequent earthworks. Personally hand-troweling the entire perimeter of the effigy, he exposed the exact layout of the original earthwork. The serpent shape was clearly outlined on the ground by a layer of clay and ash. In some spots— especially where the serpent wound across the steeper portions of the hill— the ancient architects had built up a basal layer of stones to keep their handi- work from washing away. They then built up the snake's body with a yellow clay fill and capped it with about a foot of organic-rich dark soil. Although Serpent Mound had been reduced by years of plowing before Putnam's arrival, the plow had not penetrated the main clay layer that comprised the heart of the earthworks.

Putnam also tested the nearby conical mounds and the habitation site on the ridge top. He postulated two separate occupations here, the earlier group being responsible for erecting the gigantic serpent alignment and burial mounds. The subsequent, superimposed occupation was characterized by a different kind of material culture.

After restoring each site to its original contours, Putnam converted the area into a public park that Harvard's Peabody Museum operated for a couple of years. But the long-distance administrative arrangement proved to be less than successful, and Serpent Mound was deeded to the Ohio Historical Soci- ety in 1900. Ninety-one years would pass before further excavations were undertaken at Serpent Mound.

Dating the Serpent Mound

No single, satisfying answer explains Serpent Mound to us. In fact, it is diffi- cult even to be sure when it was built. Serpent Mound does not contain human burials, and no artifacts have been found in the fill. This means that the serpent's construction phase could not be directly cross-dated or associ- ated with any particular culture.

E. F. Greenman was the first to classify the artifacts that Putnam had exca- vated in 1887, ascribing the older materials to the Adena (early Woodland) cul- ture. Noted archaeologist James B. Griffin later confirmed this assessment for

Frederic Ward Putnam, early excavator of Serpent Mound

the lower component of the village site but noted that Putnam had recovered Fort Ancient (late Prehistoric) artifacts from the upper levels of the habitation site.

Accordingly, a consensus arose over the years that the Great Serpent Mound is probably coeval with the conical (Adena-age) burial mounds nearby. Assuming that these mounds were built by the same people who constructed Serpent Mound, the snake effigy could be ascribed to the Adena builders, sometime during the first millennium B.C. But making this connection required a considerable leap of faith.

Testing the Adena Hypothesis

Because no excavations had been undertaken since 1887, archaeologists could not apply the newer research tools available to late-twentieth-century archaeologists. Then, in 1985, Robert Fletcher and Terry Cameron visited Serpent Mound to assess the possibility of astronomical alignments. They needed an

This map, prepared by Robert Fletcher and Terry Cameron, shows the probable locations of Putnam's trenches (1880s) and the 1991 excavations at Serpent Mound.

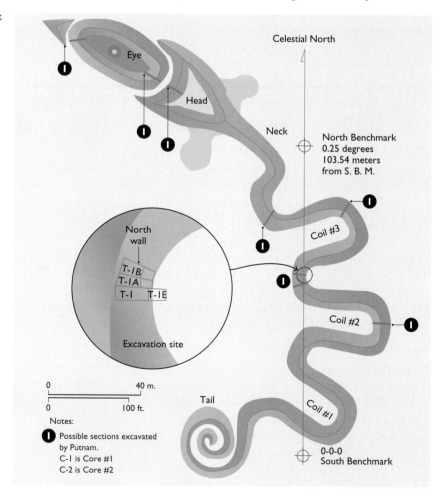

Celestial North

Eye

Head

Neck

North Benchmark
0.25 degrees
103.54 meters
from S. B. M.

Coil #3

North
wall

T-1B
T-1A
T-1   T-1E

Excavation site

Coil #2

0   40 m.

0   100 ft.

Notes:

Possible sections excavated by Putnam.
C-1 is Core #1
C-2 is Core #2

Tail

Coil #1

0-0-0
South Benchmark

accurate site map in order to take precise measurements, and they produced one the next year.

Bothered by the imprecise ages ascribed to the site, Fletcher and Cameron organized a limited excavation effort in late July 1991. They knew that Putnam had found a clay-ash-stone layer underlying the earthworks, and they thought samples of this might be datable using the new AMS (Accelerator Mass Spectrometric) approach to radiocarbon dating, which requires only miniscule samples of charcoal.

From the outset they were determined to disturb as little of the original embankment as possible. Although Putnam's notes and photographs did not permit them to pinpoint exact locations of the 1887 trenches, there were obviously depressed areas along the serpent's body that suggested relatively recent backfilling. Subsequent core sampling confirmed the location of one of Putnam's old trenches, which they reopened and extended until they encountered intact prehistoric mound stratigraphy. They then cleaned, photographed, and recorded the resulting stratigraphic profiles.

Several stone artifacts and potsherds were recovered in the 1991 excavations, but given Putnam's comments about the near-total absence of cultural materials recovered in his 1887 excavations, these probably were inclusions in the fill that Putnam used to restore the original mound contours. Thus, although we know that the recovered ceramics were manufactured between A.D. 350 and 950, their relationship to specific episodes of mound construction remains unclear.

Three radiocarbon samples were taken from the 1991 excavation profile. One, dating to about 1000 B.C., dates the original (premound) surface upon which the Serpent Mound was built. The two other samples, of oak charcoal and ash, were removed from still-intact basket-loaded sediments carried in to create the body of Serpent Mound. Radiocarbon analysis yielded the same age estimate for each sample: A.D. 1000–1140.

These estimates are derived from charcoal included in the mound fill, and they provide a minimum age estimate for the construction of Serpent Mound. The earthworks could not have been constructed before that period, but they could have been constructed later. We know from historical documentation that Serpent Mound hosted a mature stand of old-growth forest when Squier and Davis surveyed it in 1846.

A Fort Ancient Icon?

The new dates strongly suggest that Serpent Mound was constructed by people of the Fort Ancient tradition (A.D. 900–1650). This well-known culture, which occupied the middle Ohio River Valley from eastern Indiana to western West Virginia, is characterized by village settlements, subsistence based on cultivating maize and other crops, distinctive ceramics, and, at least during the later phases, stylistic connections with widespread Mississippian

These marine shell gorgets (neckpieces) show variations on the coiled rattlesnake motif.

In parts of the upper Mississippi Valley, on high ground bypassed by the Ice Age glaciers, some ancient Americans erected huge earthen likenesses of lynx, panther, bison, water birds, eagles, lizards, and turtles. Although they also built more conventional mounds in conical and linear configurations, these people are best known for their virtuoso construction of immense animal-shaped effigy mounds. Lacking a better term, archaeologists speak of the "Effigy Mound Culture" to denote this great Native American artistic tradition. Although many Native American cultures erected earthen mounds, the effigy mounds still capture the imagination, and archaeologists still puzzle over their meaning.

Perhaps there were once as many as ten thousand such mounds, but urban expansion and intensive agriculture have erased all but a handful. Effigy Mounds National Monument in northeastern Iowa has two hundred mound sites, including twenty-six animal effigy figures. Some are monumental in size. Great Bear Mound is 70 feet (21 meters) across at the shoulders, reaching 137 feet (42 meters) in length. But perhaps most impressive are the huge birds and the arc of thirteen Marching Bears

Why were these effigies built? Why here? What do they mean?

Early European visitors took them as emblematic of some "vanished race." Then, for a while, the effigies were considered to be an exotic form of burial mound, until excavation proved that, like Serpent Mound, these alleged treasure houses were mostly empty.

Today some archaeologists think that the effigies served as gigantic territorial markers preventing potential competition and conflict over the same resources. Others point out that the effigy mounds were built in distinctive groups, ranging from two or three mounds to more than a hundred. While several conical or linear mounds in a group might contain human burials, those depicting an animal shape are almost always empty, containing neither human remains nor artifacts. This suggests to some that the effigies defined sacred ceremonial ground rather than mortuary areas. Or maybe they were designators for meeting places before or after the winter break-up into smaller groups.

Perhaps it was the effigy shape itself that held the greatest significance. Some think that the animal effigies might represent clan totems, symbols depicting related family groups. Perhaps these monumental ground figures were an attempt, through ritual means, to connect with specific animal spirits, to ensure a consistent and regular food supply. To some investigators, the lack of artifacts in the mounds suggests that it may have been the actual building of the mounds that was important.

elements. (Ironically, although this tradition is named after the Fort Ancient site on Ohio's Little Miami River, that site is not typical of the Fort Ancient tradition.)

Assigning Serpent Mound to the late Prehistoric period has major stylistic implications. Serpent depictions appear in great numbers about this time, particularly the "serpent eye" motif on rattlesnake shell gorgets, occurring throughout Mississippian sites and also in graves of the late Fort Ancient tradition.

Further Reading

R. C. Glotzhober and Bradley T. Lepper have written a first-rate discussion in *Serpent Mound: Ohio's Enigmatic Effigy Mound* (Columbus: Ohio Historical Society, 1994). Classic presentations include *Ancient Monuments of the Mississippi Valley* by E. G. Squier

Still others emphasize the importance of the heavens in traditional Native American cosmology. Many American Indian groups paid close attention to what was overhead, intertwining their cultural rhythms with the perpetual cycles of sun, moon, planets, and stars. They observed eclipses and the conjunctions of planets, devised calendars for festivals, and established dates for planting. All this was vital knowledge. Today's astronomers capture such information in books and scientific journals. Ancient Americans may have done the same thing through myths, rituals and festivals, symbolic architecture, dance, and costume.

Is it significant that on early spring evenings the Big Dipper is located precisely over the top of this arc? Does it matter that during late summer you can see the Big Dipper exactly over the bottom position of the effigies? Could it be that the earthen Marching Bears represent the march of the Big Dipper around Polaris, the north star? Some astronomers think so.

Whatever they meant to the people who made them, the enigmatic effigy mounds today are an archaeological Rorschach test against which to project one's personal beliefs about the past and those who lived it.

Aerial photograph of the Great Bear Marching Bear group, containing ten "bears," three "birds," and two linear mounds. These dynamic representations, probably built between A.D. 600 and 1300, are one of the largest groups of effigy mounds still surviving.

and E. H. Davis (originally published in *Smithsonian Contributions to Knowledge* [1851] and recently republished by the Smithsonian Institution Press) and E. F. Greenman's *Guide to Serpent Mound* (Columbus: Ohio Historical Society, 1934).

Further Visiting

Serpent Mound (Locust Grove, OH; 4 miles [6 kilometers] northwest on SR 73) has museum facilities and a viewing tower located on site.

TEN

Mesa Verde

A.D. 600 – 1300

Ancestral Pueblo culture

in Colorado

When millions of Americans think of archaeology, they think of Mesa Verde in Colorado. The name—Spanish for "green table"—refers to the region's comparatively flat tablelands, heavily forested with juniper and piñon trees. The oldest archaeological national park in the United States, Mesa Verde encompasses 80 square miles (200 square kilometers), rising 1,800 to 2,000 feet (550 to 610 meters) above the north-side valley, sloping down to the cliffs bordering the Mancos River Canyon on the south. A score of spacious canyons cut into Mesa Verde. Many years ago people constructed some of the world's most renowned archaeological sites in the shelter provided by the hundreds of alcoves eroded into these cliffs. Cliff Palace is the most famous ruin at Mesa Verde, something of an international celebrity.

Discovering Cliff Palace

In the early 1880s the Wetherill family moved to the Four Corners area, where the states of New Mexico, Colorado, Utah, and Arizona meet. The Wetherills were Quakers, and their Alamo Ranch near Mancos, Colorado, soon became home to the starved or infirm Ute Indians spurned by other white settlers. The Utes expressed their gratitude by allowing the Wetherills to run their cattle on remote tribal land to the west—backcountry that was off-limits to other Anglos. As he ran cattle across the massive tablelands of Mesa Verde, Richard Wetherill and his brothers found several small ruins tucked into sandstone overhangs. The Wetherill boys continued to poke around, often finding what they termed "cliff dweller" artifacts.

As the years passed, rumors surfaced of a huge ruin high up in the rocks of Cliff Canyon, far beyond where Anglos had ever gone. Although Wetherill asked his Ute friends to take him there, they refused to disturb the spirits of those who had once lived there. Then, in late 1888, Richard Wetherill and his brother-in-law Charlie Mason were driving strays well up into Cliff Canyon. Two dozen miles from home, they rode along the Mesa Verde rimrock. The sun was fading, and the drovers looked for someplace to spend a long, wintry night.

Richard Wetherill rediscovered Cliff Palace in 1888.

Engraving (1893) of Gustaf
Nordenskiöld climbing up a
fortified cliff at Mesa Verde.

Years later, novelist Willa Cather described Wetherill's discovery in *The Professor's House* (1925):

> In stopping to take breath, I happened to glance up at the canyon
> wall. I wish I could tell you what I saw there, just as I saw it, on that
> first morning, through the veil of lightly falling snow. Far up above me,
> a thousand feet or so, set in a great cavern in a cliff, I saw a little city
> of stone, asleep. It was as still as sculpture—and something like that. It
> all hung together, seemed to have a kind of composition: pale little
> houses of stone close to one another, perched on top of each other,
> with flat roofs, narrow windows, straight walls, and in the middle of
> the group, a round tower. . . .
>
> That village sat looking down into the canyon with the calmness
> of eternity. The falling snow-flakes, sprinkling the piñons, gave a spe-
> cial kind of solemnity. I can't describe it. It was more like sculpture
> than anything else. I knew at once that I had come upon the city of
> some extinct civilization, hidden away in this inaccessible mesa for
> centuries, preserved in the dry air and almost perpetual sunlight like a
> fly in amber, guarded by the cliff and the river and the desert.

Leaving the cattle to fend for themselves, Wetherill and Mason cobbled together a makeshift ladder and clambered over the rimrock. As they poked about the ruin, they found everyday implements discarded as though their owners had just stepped away for a moment: a stone axe still lashed to its wooden handle, unbroken pottery mugs, and water jars.

Wetherill called the place the Cliff Palace, a name today known around the world. Cliff Palace is the largest, most celebrated, and best preserved of all the cliff dwellings. At the time, Richard Wetherill was a full-time cowboy and part-time pothunter, always on the lookout for whatever relics he could dig up and sell. But his Mesa Verde discoveries would soon convert him from a relic-collecting hobbyist to an aspiring professional archaeologist.

During the summer of 1891, Wetherill received invaluable on-the-job training from Baron Gustaf Nordenskiöld, a Swedish geologist who taught him the basics of European archaeological technique: substituting a hand trowel for a shovel, making both fieldnotes and photographic records. Always the cagey businessman, Wetherill also realized that the value of his collections would be established not just by the quality of the artifacts but also by his ability to document every step of the work. Nordenskiöld's lavishly illustrated *The Cliff Dwellers of Mesa Verde*, published in 1893, would soon tell the world about Cliff House and the cliff dwellers who lived there.

A New Light on the First Farmers

Wetherill's Mesa Verde collections made their way to the Chicago World Columbian Exposition that same year, and his frontier lexicon remains with us today. Not only did Wetherill name Cliff House, but he was the first to use

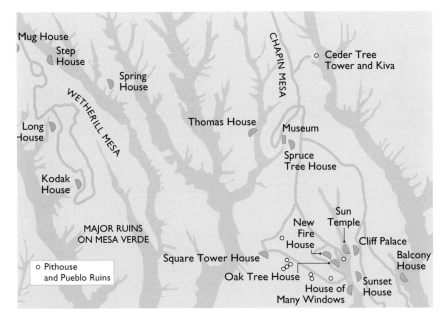

Map showing the major ruins at Mesa Verde

Mug House
Step House
Spring House
Long House
Kodak House
Thomas House
CHAPIN MESA
Ceder Tree Tower and Kiva
Museum
Spruce Tree House
WETHERILL MESA
Sun Temple
New Fire House
Square Tower House
Oak Tree House
House of Many Windows
Cliff Palace
Balcony House
Sunset House

MAJOR RUINS ON MESA VERDE

Pithouse and Pueblo Ruins

the terms "Anasazi," "cliff dweller," and "basketmaker" in reference to the remarkable ancient people of Mesa Verde. Wetherill's terminology formed the basis for the so-called Pecos classification, which survives in modified form to this day.

In this scheme, the Anasazi tradition began with a hypothetical "Basketmaker I" stage, allowing for then-undiscovered nonagricultural ancestors of the Basketmaker people. Recently, however, archaeologists have accumulated ample evidence for Archaic hunter-gatherers in the Southwest, and the term Basketmaker I is no longer used.

ANASAZI OR ANCESTRAL PUEBLO?

When Richard Wetherill asked his Navajo friends who had built the wonderful buildings at Mesa Verde, they replied "Anasazi"—the "ancient enemies." Wetherill started using the name Anasazi for the architects of the spectacular sites in the Four Corners area.

Years later, archaeologist A. V. Kidder made Anasazi the standard term for the major cultural tradition ancestral to modern Pueblo people of the American Southwest. Recognizing the need to subdivide the overall Anasazi tradition, Kidder convened a conference in 1927 at Pecos, New Mexico, where he had been conducting long-term excavations. This first Pecos Conference, as it became known, produced a classification of eight sequential periods for the Anasazi area:

Basketmaker I: postulated nonagricultural population; used atlatl and did not make pottery

Basketmaker II (or simply Basketmaker): originally thought to date A.D. 1–450, now believed to date 500/100 B.C.–A.D. 500; agricultural, non-pottery making, used atlatl

Basketmaker III (or late Basketmaker): 450–700/750; lived in pit and slab houses; did not practice cranial deformation

Pueblo I (or proto-Pueblo): 700/750–900; villagers living in rectangular, true-masonry houses; first to practice cranial deformation; introduced neck-corrugated pottery

Pueblo II: 900–1050; widespread proliferation of small villages; ceramics characterized by elaborate corrugations over entire surface of vessel

Pueblo III (or Great Pueblo Period): 1050–1300; construction of large communities; intensive local specializations and development of the arts

Pueblo IV (or proto-Historic): 1300–1600; populated area contracted; corrugated pottery gradually disappeared; general decline from previous stage

Pueblo V (or Historic): 1600–present; modern Pueblo people

The Pecos Classification was proposed before tree-ring dates became widely available. Before long it became clear that cultural developments did not take place at a uniform rate throughout Anasazi territory. In this sense, the categories are probably better considered to be successive stages rather than discrete periods. Although modified considerably, the Pecos classification brought order out of typological chaos. It remains useful today, particularly in the Four Corners area.

In recent years many Pueblo people have expressed concern over continued use of the term "Anasazi." Why, they ask, should their ancestors be known by a derogatory non-Pueblo (Navajo) term? Although a number of substitutes have been suggested, many archaeologists today prefer the descriptive term "ancestral Pueblo" to "Anasazi." The matter is far from settled, and here we use both terms more or less interchangeably.

Basketmaker plaited basketry with herringbone design from Grand Gulch, Utah. When found by Richard Wetherill, it was stuffed to the brim with red and yellow corn kernels.

The Pecos classification then posited a model of gradual change. For decades, southwestern archaeologists saw Mesa Verde as representing the long-term, progressive unfolding of ancestral Pueblo (Anasazi) culture. The early Archaic people lived in small, nomadic groups. As they gradually picked up new innovations—farming, pottery, and more permanent masonry architecture—the overall population size increased, and people began living in larger settlements, eventually approximating those of their historic-period Pueblo descendants.

But recent research has demolished this uniform, gradual view of early Anasazi prehistory. The Dolores Archaeological Project—one of the single largest archaeological projects in the history of the United States—revised the chronology for the Mesa Verde area. In particular, the Basketmaker II period was significantly lengthened, now beginning about 500–1000 B.C. and lasting as late as A.D. 500. In this view, Basketmaker II lasts as long as the rest of the entire Anasazi/Pueblo sequence, maybe even longer.

The earliest Basketmaker II people carried over much from their Archaic ancestors: use of the atlatl (spear thrower); collecting wild seeds to be processed on grinding stones (small manos and metates), small community size with little investment in building permanent housing. The importance of agriculture in the early part of Basketmaker II remains unclear.

Characteristic geometric designs on the interior of black-on-white Anasazi bowls.

It now seems likely that there was a break between late Archaic and Basketmaker occupations in the Mesa Verde area. R. G. Matson has argued that some Basketmaker II sites, including the famous Grand Gulch occupations, might well result from the physical migration of people from southern Arizona, where archaeologists have recently excavated very early agricultural villages dating from about 800 B.C. to A.D. 100. As discussed in Chapter 12, these early Arizona farmers might be the direct ancestors of the Basketmaker II farmers. Matson thinks that perhaps there is a lag in the northern introduction of maize cultivation because the new domesticates needed time to adapt to different growing conditions there, especially seasonal rainfall patterns and the shorter growing season.

During the later (A.D.) part of the Basketmaker II period, people had become heavily dependent on floodwater farming. Basketmakers cultivated maize (corn), squash, and beans, upland plants initially domesticated in Middle America. Such domestication effectively converted a wild species to one dependent upon human intervention for survival. Full-blown agricultural domestication took place when practices like weeding, irrigation, and plowing created new opportunities for plant evolution.

Not only were plant genetics modified, but human behaviors were affected as well. Becoming a farmer meant taking care of cultivated plants, although the adoption of farming did not necessarily mean that only domestic species were eaten. During the historic period, the Pueblo and O'odham (Pima) peoples of Arizona, for instance, relied on domesticated foods for perhaps 50 to 70 percent of their diet. In some years, they ate little or no domesticates; other years, they lived almost exclusively on their domesticated crops. When Hopi crops failed—as they did from time to time—the Pueblos broke up into smaller family foraging groups, which were quite capable of living off naturally available foodstuffs.

Recent evidence suggests that during most years, the late Basketmaker II farmers probably got the majority of their calories from maize. But the region maintained a sufficiently low population that foraging remained a potential fallback strategy when crops failed.

During the subsequent Basketmaker III period, people began making pottery, mostly gray and black-on-white ceramics. The Dolores Archaeological Project has demonstrated that Basketmaker III farmers in the greater Mesa Verde area lived in dispersed single family homesteads. They built large pit structures, with nearby above- and below-ground storage buildings. They dug numerous lined storage containers called cists both inside and outside their houses.

Early Puebloans

Across the Colorado Plateau, the Anasazi entered the Pueblo I phase (A.D. 750–900), marking the transition from pit houses to surface dwellings. Instead of storing their surplus supplies in cists and living strictly in pit houses as in Basketmaker times, these ancestral Pueblo people built a series of storage rooms behind the pit structures.

The classic Pueblo I residential complex is called the Prudden Unit in honor of early-twentieth-century archaeologist T. Mitchell Prudden. It consists of a large deep pit structure sometimes called a protokiva, along with a few closely associated surface living and storage rooms. Ramadas (open porches) stand in front of the surface rooms, which people constructed by weaving branches through uprights, then plastering these walls with mud. As these wattle-and-daub walls were connected to one another, large habitation spaces were created above ground, presaging the Pueblo architectural style that continues to this day in the American Southwest.

The sunken pit structure appears to be the primary living room. Although apparently used for domestic activities, these pit structures also commonly exhibited ritual features, such as a sipapu (an entrance to the spiritual world), implying a gradual change in function away from purely secular domestic activities. Some researchers believe that two or three nuclear families used the surface rooms separately and shared the pit structure. But Ricky Lightfoot has recently argued that a single household—a nuclear family or small extended family—might have used both the pit structure and the associated surface-room suites.

Prudden Units sometimes occur by themselves in a fully dispersed settlement pattern. But when the regional population aggregated into hamlets or villages, the Prudden Units were joined together, forming side-by-side room blocks. A Pueblo I village might contain several of these multiunit room blocks, scattered within sight of one other.

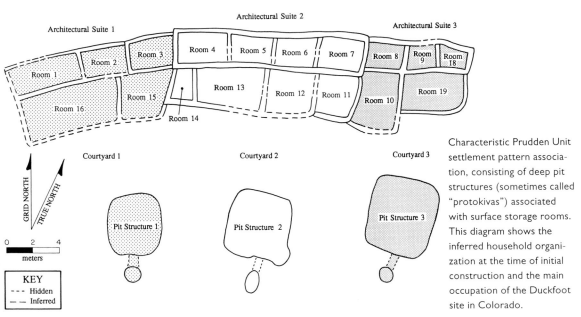

Characteristic Prudden Unit settlement pattern association, consisting of deep pit structures (sometimes called "protokivas") associated with surface storage rooms. This diagram shows the inferred household organization at the time of initial construction and the main occupation of the Duckfoot site in Colorado.

The Dolores project identified several "great pit structures"—versions of the regular pit structures used residentially, but without associated surface living and storage rooms. These are ritually specialized structures, with elaborate floor vaults and complex sipapus, probably serving the ritual needs of multiple household alliances.

There is considerable variability in Mesa Verde settlement patterns during the Pueblo I period. Some areas retained dispersed communities like those of Basketmaker III times. But elsewhere Pueblo I settlements occured as true villages with more than fifty to one hundred people living in close proximity. Some Pueblo I villages are actually larger than the villages of the Pueblo III period, the era of the better-known settlements like Cliff Palace. Some archaeologists believe that the Pueblo I villages at Mesa Verde may represent the largest overall population of the area.

It does seem clear, however, that these large Pueblo I villages were short-lived, reflecting what Lipe and Matson have called a "boom and bust" cycle of build-up and collapse. Perhaps early farming could not be sustained ecologically over long periods of time, or maybe the Anasazi social fabric was incapable of integrating larger congregations of people.

Archaeologist Douglas Osborne with a sample of artifacts recovered from Mesa Verde

Overall, Pueblo I population density in the Mesa Verde area increased during the late A.D. 700s, then declined in the early 800s. By the mid-800s, the population rose very rapidly, almost certainly due to immigration into the Mesa Verde area. Larger villages of up to one thousand inhabitants were established at this time. Population levels then plummeted, and by 900 the area was virtually unoccupied, although a few families seem to have lingered on. Some believe that the Mesa Verde population drifted southward, augmenting the population at Chaco Canyon in New Mexico.

By the early 1100s, the greater Mesa Verde area has been drawn into a Chaco-related sphere of interaction. Between 1080 and 1130, many long-standing Mesa Verde communities built multistoried Great Houses in the Chaco Canyon style, providing a form of "public architecture" for the Pueblo I citizenry, most of whom still lived in Prudden Units dispersed across the landscape in farmsteads or small hamlets. Great Kivas are commonly associated with the Great Houses.

The middle 1100s remain elusive. During these few decades, the northern San Juan area was hit by the worst drought of the Basketmaker-Pueblo period. People stopped building Chacoan Great Houses about 1125–1130. Cannibalism has been documented for about a dozen drought-era sites (1125–1150), after which sites of any kind become rare. Perhaps people abandoned the area entirely, although there is no evidence of a migration into neighboring regions. Or maybe people were just barely hanging on, no longer raising buildings and thus generating few beams for tree-ring dating.

The Great Pueblo Period

During the so-called Great Pueblo period (Pueblo III, 1150–1300), population increased rapidly. The 1200s probably saw the peak population of the entire ancestral Pueblo sequence. People aggregated into Pueblo III communities with more than one thousand inhabitants. Great Kivas or distinctive tri-wall structures provided communal meeting areas and places for conducting secret rituals.

People still lived in Prudden Units (a few spatially associated surface rooms) throughout the Pueblo II and III periods. Pit structures during these later periods become round. Archaeologists call them "kivas" (although Bill Lipe thinks they were a kind of "family room-with-a-shrine-in-it"). Although Prudden Units disappeared in the late 1200s, the rounded semisubterranean form, sipapu, and floor vault would survive as more ritually specialized "true" kivas of the Pueblo IV and historic periods to the south and southeast, where Pueblo people continued to live after depopulation of the Four Corners area.

Most of the cliff dwellings were built between 1230 and 1240, when people moved from open mesa-top communities down into more easily defended shelters and ledges along the canyon walls. During this period, people moved

into Cliff Palace, which had 220 rooms and 23 distinctive keyhole-shaped kivas. Many of the thirty-some cliff dwellings at Mesa Verde had more than five hundred rooms, with perhaps six hundred to eight hundred people sharing close, if well-protected quarters. Sites were also built along canyon rims and on the talus slopes below the rim. The rule was simple: if shelter was available, Pueblo III housing would be built inside.

Cliff dwellers still farmed the land on the mesa tops, creating some problems due to the difficult access between residential areas and the outlying agricultural fields. Because it was hard to protect the faraway fields from human and nonhuman predators, Pueblo farmers often built field houses for temporary shelter.

Unfortunately, many of the spectacular sites at Mesa Verde were excavated before the development of modern archaeological techniques. Although spectacularly well-preserved artifacts and architecture survived these roughshod early excavations, much of the contextual information necessary to reconstruct the social organization of the Mesa Verde people was destroyed. In part to remedy this problem, Arthur Rohn set out in the 1960s to launch painstaking excavations of Mug House, a ninety-four-room cliff dwelling built during the 1100s. Rohn demonstrated that the earliest inhabitants of Mug House constructed a series of rooms with an associated kiva and courtyard. Through time, the settlement evolved into four distinct household clusters. These eventually merged into a single Pueblo settlement during the latest period of occupation, but the northern and southern halves were separated by a wall with no direct internal access between them. Perhaps they represent different social units, as the five northern kivas are quite similar to each other but differ from the southern kivas.

The human population of the Mesa Verde area began dropping after 1250. Full-scale evacuation took place during the late 1260s and 1270s, with the area vacated by the late 1280s or early 1290s. By 1300 the ancestral Pueblo people had permanently abandoned the entire Four Corners region, an area of more than 23,000 square miles (60,000 square kilometers). Explaining the Puebloan abandonment of the Four Corners has been one of the classic challenges of American archaeology.

The problem is simple: During the mid-1200s the Pueblo population numbered at least ten thousand people. Five decades later, these people had disappeared. Archaeologists still debate what happened. Lipe identifies a number of factors that may have "pushed" Puebloans out of the Four Corners area, fingering drought and disease, warfare and factionalism as possible causes for the abandonment.

Without doubt, a major drought stuck between 1276 and 1299, but its consequences remain uncertain. Recent paleoenvironmental evidence suggests that, even though the "Great Drought" must have adversely affected Pueblo farmers, they still could probably have grown sufficient crops to feed the populace.

Warfare was also a factor. Not only were individuals and communities at risk, but warfare may also have taxed the social system itself. As people grouped together for communal defense, family households were no longer free to move away from strife within their own communities. Lipe suggests that intracommunity friction could have made migration more appealing. But persistent warfare also meant that once abandonment began, individual families or small community isolates were not free to linger on their own, and depopulation was complete.

In addition to the "push" factors, decisions to migrate may also have been hastened by the "pull" of apparently more favorable areas to the south. During

Jar with flattened corrugations and indented spiral band from Mesa Verde

the late 1200s, the Rio Grande and western Pueblo areas enjoyed more reliable summer rainfall than did Mesa Verde. The religious practices of these two "recipient" areas may also have been attractive to the outmigrating Mesa Verde people, who left behind much of their own religious symbolism and practice.

Whatever the precise mix of "push" and "pull" factors, it is clear that 1250–1350 was a time of large-scale demographic shifts, coupled with dynamic change in Pueblo social organization, religion, and symbolism. Perhaps the Four Corner abandonment was merely the culmination of long-term boom-and-bust cycling, which overused the land in a futile attempt to support the greatly enlarged Anasazi population. In any case, archaeologists dating back to Richard Wetherill have clearly recognized the close affinity between the ancient ruins at Mesa Verde and the pueblos built by modern Pueblo Indian people at Acoma, Zuni, Hopi, Taos, and elsewhere in the American Southwest.

Further Reading

The best available summary is Linda S. Cordell's *Prehistory of the Southwest*, second edition (Orlando, FL: Academic Press, 1997). David Grant Noble has written a non-technical introduction, *Ancient Ruins of the Southwest: An Archaeological Guide*, revised edition. (Flagstaff, AZ: Northland Publishing, 1991). I also recommend *The Anasazi* by J. J. Brody (New York: Rizzoli, 1990); *Understanding the Anasazi of Mesa Verde and Hovenweep*, edited by David Grant Noble (Santa Fe: School of American Research, 1985); and *Wetherill Mesa Excavations: Mug House, Mesa Verde National Park—Colorado* Arthur Rohn (Washington, DC: National Park Service Archeological Research Series, 1971).

Early Anasazi agriculture is presented in *The Origins of Southwestern Agriculture* by R. G. Matson (Tucson: University of Arizona Press, 1991); *The Origins of Agriculture: An International Perspective*, edited by C. Wesley Cowan and Patty Jo Watson (Washington, DC: Smithsonian Institution Press, 1992); and "Anasazi Origins: Recent Research on Basketmaker II" by R. G. Matson and Karen M. Dohm, *The Kiva* 60, no. 2 (1994): 159–64.

Further Visiting

Mesa Verde National Park (southwestern CO) is one of the premier archaeological refuges in this country. During the summer season, visitors can also drive along a 12-mile (19-kilometer) access road to Wetherill Mesa, where two cliff dwellings and four earlier mesa-top villages are open.

ELEVEN

Pueblo Bonito

A.D. 860–1150

Ancestral Pueblo culture

at Chaco Canyon, New Mexico

Lieutenant James B. Simpson, surveyor for the U.S. Army in 1848, could not believe his eyes when he came upon the Chaco Canyon ruins. Nothing in his past experience prepared him for the sight of hundreds of contiguous rooms of beautifully shaped and coursed stonework, three and four stories high, forming huge sweeping arcs. More than a dozen Great Houses, each containing hundreds of rooms, had grown up along a 9-mile (14-kilometer) stretch of the canyon that Simpson called Chaco—probably an Anglicized version of the Hispanicized Navajo name for Chacra Mesa (*Tzak aih*, meaning "white string of rocks").

Simpson refused to believe that the ancestors of modern Pueblo Indians could possibly have constructed these massive Chacoan towns. Dismissing Pueblo building techniques as too crude, Simpson called up the ancient Toltec tribes of Mexico as the architects of Pueblo Bonito (Spanish for "beautiful town") and the other Great Houses. The years would produce other candidates, including the Aztecs and even the Romans.

Richard Wetherill, the cowboy-archaeologist who "discovered" Mesa Verde in Colorado (see Chapter 10), first visited Chaco Canyon in 1895. Despite his extensive experience in Four Corners archaeology,

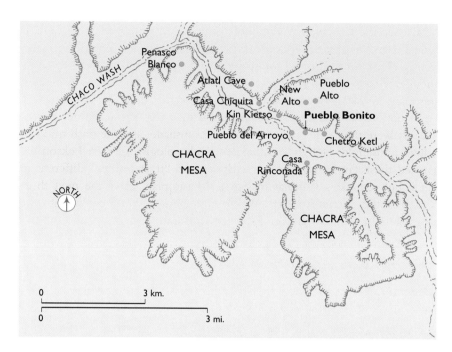

128

Wetherill was astonished by the Chaco sites. He was immediately drawn to Pueblo Bonito, which had once stood five stories high. Its ruins covered nearly 3 acres (1.5 hectares). Nearby Chetro Ketl was almost as large, and not far away were a half dozen other huge ruined buildings. The sandstone masonry was finer than anything he had seen in the Pueblo world.

Unlike most Anasazi buildings, which lasted only a matter of decades, the Chacoan Great Houses were built to endure. When Richard Wetherill came to excavate Pueblo Bonito—seven centuries after the site had been abandoned—he found that entire room blocks were still usable, ceilings and walls perfectly intact. Enlisting the support of New York's American Museum of Natural History, Wetherill literally anchored his camp, headquarters, and trading post to the still-sturdy rear wall of Pueblo Bonito.

Although Wetherill did not know it, Pueblo Bonito was the largest edifice of its time in North America. It could house a thousand people. In fact, Pueblo Bonito had been the largest building in America until 1882—thirteen years before Wetherill's arrival in Chaco Canyon—when the Spanish Flats apartments were erected in New York City.

Evolving Chacoan Architecture

Pueblo Bonito was not built overnight. Its architectural origins can be traced to the earliest Chacoan pit houses, dating perhaps from A.D. 500. Their semi-subterranean rooms were roofed with thick layers of dirt over wooden frames. As we saw in Chapter 10, people had used dwellings like this for hundreds of years, even before the development of pottery or agriculture.

By 800 the Chacoans were building small blocks of three or more individual rooms near each pit house; these probably served as home to a single family. These above-ground rooms were initially storerooms, but eventually they came to be used for corn-grinding, cooking, and warm-weather sleeping. Early Anasazi masonry consisted either of wattle and daub (mud and stick constructions) or rough-shaped stones laid in a bed of thick mud mortar. Nearby were the *ramada* (a pole and brush sunshade) and the trash midden, where household discards accumulated. Archaeologist Stephen Lekson has termed these family homesteads early Anasazi "tract homes," different in detail but remarkably similar in size and basic plan. Over the next three centuries, these small Ancestral Pueblo unit houses would evolve into the singular architectural unit known as Pueblo Bonito.

Tree-Ring Dating at Pueblo Bonito

A century of concerted archaeological research has produced an extraordinarily detailed tree-ring sequence to unravel the architectural complexities

This aerial photograph of Pueblo Bonito was taken by Charles Lindbergh in 1929. "Threatening Rock," a great 30-ton stone monolith can be seen towering over the site. Ancient Anasazi builders had tried to stabilize it with pine pole props and broad stone terraces. For years, the National Park Service precisely monitored its movements until, during the afternoon of 21 January 1941, Threatening Rock buckled and fell, crushing 65 rooms along the north wall of Pueblo Bonito.

in Chaco Canyon. Wetherill excavated Pueblo Bonito for four field seasons, recovering plenty of preserved wooden elements such as roof parts, lintels, pilasters, wall pegs, and kiva wainscoting. These samples were eventually turned over to A. E. Douglass for his pioneering study of the American Southwest.

During the 1920s, Neil Judd and his Smithsonian Institution team conducted exemplary excavations at Pueblo Bonito, collecting a number of tree-ring samples. Further long-term projects were undertaken by the Museum of New Mexico, the University of New Mexico, and the National Park Service; each extracted well-provenanced dendrochronological samples. By the 1980s, 218 tree-ring dates were available from Pueblo Bonito. An additional 390 dates since then make Pueblo Bonito one of the best-dated sites in the world.

But what do all these tree-ring dates actually tell us? At a minimum, dendrochronology says this about Pueblo Bonito: The building was erected between A.D. 860 and 1129. By most archaeological standards, this degree of chronological precision is extraordinary. But when examined in detail, the dating of Pueblo Bonito is considerably more complex than this simple time-span might suggest.

We know that the site and buildings of Pueblo Bonito were used, off and on, for more than 1,300 years. Earlier (Basketmaker III or Pueblo I) structures lie buried below the massive Pueblo Bonito ruins, and nineteenth- and early-twentieth-century Navajos and Euro-Americans reused the building for

Major construction episodes at Pueblo Bonito, based on tree-ring data collected by Thomas Windes and Dabney Ford

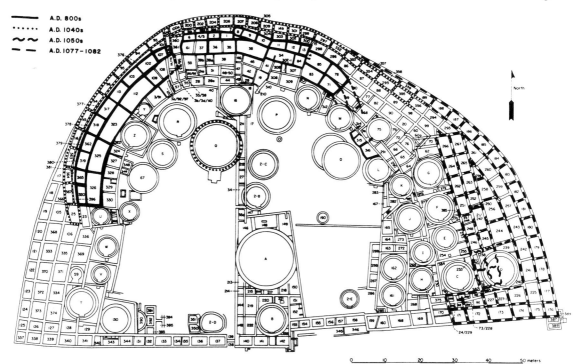

A.D. 800s
A.D. 1040s
A.D. 1050s
A.D. 1077–1082

This reconstruction of Pueblo Bonito, by Richard Schlect, shows how this five-story apartment-town may have appeared at its peak.

both storage and shelter. Wood has always been a valuable commodity in Chaco Canyon, and people have extensively mined places like Pueblo Bonito for wooden construction materials and firewood over the centuries. This extensive recycling and reworking of wooden elements has created an incredibly complex dendrochronological record. The Chaco Wood Project was initiated in 1985 to address such complexities. Thomas Windes and Dabney Ford have recently produced a new, fine-grained construction sequence for Pueblo Bonito.

The first major burst of construction activities took place at Pueblo Bonito in 860–862, followed by another in 891. By the late 800s roughly 145 rooms had been constructed in a curving arc around a plaza with six kivas. Even at this early stage, the distinctive curvature of Pueblo Bonito was already established.

Although these units followed the generalized Chacoan pattern, they

Like many of archaeology's dating techniques, tree-ring dating or dendrochronology was developed by a non-archaeologist. A. E. Douglass, an astronomer by training, knew that because each ring in a tree represents a single year, it is, in theory, a simple matter to determine the age of a newly felled tree: count the rings. Douglass took this relatively common knowledge one step further, reasoning that because tree rings vary in size, they may preserve information about the environment in which individual trees grew. Because environmental patterning affects all the trees maturing in the same place at the same time, these irregular patterns of tree growth (that is, ring width) should overlap to create a long-term chronological sequence.

Douglass began his tree-ring chronology with living specimens. He would examine a stump or a core from a living tree, count the rings, then overlap this sequence with a somewhat older set of rings from another tree. But dead trees and surface snags went back only five hundred years or so. Further back, dendrochronology had to rely on the prehistoric record.

Fortunately, Douglass was working in the American Southwest, where arid conditions enhance preservation. By turning to archaeological ruins, Douglass began mining a vast quarry of tree-ring data. Sampling ancient beams and supports, he slowly constructed a prehistoric "floating chronology," spanning several centuries but not tied into the modern samples. Douglass could use his floating sequence to date various ruins relative to one another, but the hiatus between prehistoric and modern sequences meant that his chronology could not be correlated with the modern calendar.

This "gap"—the unknown span of time separating the ancient, prehistoric sequence from the known, historically grounded chronology—plagued southwestern archaeologists for years. Finally, in 1929, an expedition at Showlow, Arizona, found a specimen that neatly bridged the gap. The sequences were united, and almost overnight Douglass was able to tell southwestern archaeologists when their most important sites had been built. The cliff dwellings of Mesa Verde were erected between A.D. 1180 and 1280, Pueblo Bonito between 919 and 1130. At last dozens of southwestern sites could be accurately dated.

Since then, the dendrochronological sequence has been extended back millennia. In addition, experts are building tree-ring chronologies for many other areas, including the American Arctic, the Great Plains, Germany, Great Britain, Scandinavia, Ireland, Turkey, Japan, and Russia.

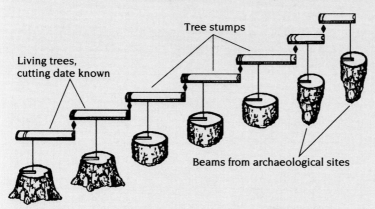

Living trees, cutting date known

Tree stumps

Beams from archaeological sites

Diagram illustrating how dendrochronologists date archaeological sites, by working from living trees to ancient beams

differed from most other Anasazi houses because they were bigger and because the rear rooms had second stories. This was probably the first multistoried architecture attempted in the American Southwest; but its early walls stood for only about a century before they crumbled, to be replaced with a new, stronger kind of stone masonry.

Another major building flurry took place in 1047–1049. As Windes and Ford point out, these earlier building episodes correspond closely with intervals of significantly increased precipitation in the region. Assuming that additional moisture translated into greater agricultural productivity, the dendrochonological evidence suggests that Great House construction took place during episodes of above-normal food supply. Still another major addition was built in 1077–1081, during another extraordinarily wet decade. These new rooms were probably storage facilities for local horticultural surplus, or perhaps public works projects needed during periods of plenty.

Considerable wood was cut for Pueblo Bonito in 1029, after a wet period. But this wood was not actually used for new construction at Pueblo Bonito until 1049 and later, when additional trees were cut during the subsequent two decades of below-normal precipitation, presumably a period of agricultural decline in Chaco Canyon. In other words, during this relatively arid interval, the role of Pueblo Bonito shifted from local storehouse to road-related rooms, perhaps used as lodging for visitors, for storage of goods entering Chaco Canyon, or for security of valuables left behind by those departing from Chaco Canyon. If true, this suggests that by 1040 the economic and social power structure in Chaco Canyon had transcended the local resident communities and was instead tying communities together at a regional level.

All this construction took a tremendous toll on local resources. Up to 50,000 trees were harvested just to build Pueblo Bonito (to say nothing of the dozen or so other Great Houses in Chaco Canyon). In conjunction with their microchronological studies, Windes and Ford were able to establish long-term patterns of wood use in Chaco Canyon.

The earliest builders of Pueblo Bonito cut mostly cottonwoods and ponderosa pines, supplemented with occasional piñons, Douglas firs, and junipers. By the early 900s the ancestral Puebloans had apparently denuded the immediate Chaco Canyon area for construction and firewood. Not only were local tree stands depleted, but enlarging Pueblo Bonito required great quantities of long straight beams capable of spanning the 10- to 13-foot (3- to 4-meter) spaces engineered inside the Great House.

Tenth-century builders looked to ponderosa pine, spruce, and fir stands in the surrounding mountains, such as the Chuskas to the west, for timber. These forests were not harvested haphazardly. Apparently Ancestral Pueblo wood cutters carefully selected trees of uniform size and age, effectively thinning out, rather than decimating, the high-altitude forests. Current evidence suggests that ancestral Puebloans may have satisfied the demands of Chacoan construction for centuries without inflicting widespread damage on the faraway mountain environments.

Building the Chacoan Great Houses required that perhaps a quarter of a million trees be harvested—a task far easier said than done. The Anasazi woodcutters probably felled and branched the trees while still green. They must have also cut them to the desired length at this point, which probably explains the standardization of Chacoan room sizes. The average primary beam at Pueblo Bonito—about 9 inches (23 centimeters) in diameter and 18 feet (5 meters) long—weighed a little over 0.25 metric tons. The Anasazi left the logs to dry thoroughly before hauling them to Chaco Canyon. The lack of transportation scars indicates that the logs were carried, rather than dragged or rolled.

This is a stupendous feat: Think about chopping down trees in the forested mountains (using only stone tools), then carrying thousands of telephone pole-sized beams to Pueblo Bonito, perhaps 40 miles (65 kilometers) away. It is small wonder that Lynne Sebastian considers wood a major "wealth item" at Chaco Canyon.

The Chacoan Road System

During the first half-century of archaeological research in Chaco Canyon, investigators concentrated mostly on the Great Houses. But within its approximately 30 square miles (75 square kilometers), the canyon contains more than

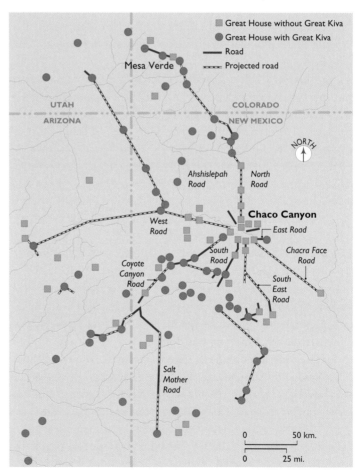

Schematic diagram of the Chaco road system, as it may have existed in A.D. 1050. Seven major road segments led into Chaco Canyon, although reconstructing specific roadways remains problematic.

2,400 archaeological sites, and as time went by, archaeologists realized the potential in studying some of the smaller and more specialized Chacoan sites.

In 1970–1971 archaeologist R. Gwinn Vivian mapped what he thought was a series of ancient Ancestral Pueblo canals in Chaco Canyon. As he began excavating, Vivian realized that this linear feature was like no canal he had ever seen. Instead of being U-shaped in profile, the Chaco "canal" was instead a deliberately flattened and carefully engineered roadway. Navajos living in Chaco Canyon and some early archaeologists had noticed portions of the ancient roadways, but they lacked the technology to trace these possibilities more than a mile or two, and their occasional speculations attracted little interest.

Vivian described his curious find to Thomas Lyons, a geologist hired to experiment with high-tech approaches to Chacoan archaeology. Comparing a set of aerial photographs taken in the 1960s with a 1930s series taken before grazing was permitted at Chaco National Monument, they saw unmistakable traces of a prehistoric road network. Even photographs of Chaco Canyon taken by famed aviator Charles Lindbergh in the 1920s clearly showed the Chacoan roads. New overflights were commissioned, and road segments were field-checked against the aerial photographs. By the late 1970s, more than 80 miles (130 kilometers) of prehistoric roads had been confidently identified in the Chaco area.

Today the length of the Chacoan road system remains controversial. Aerial photographs suggest that perhaps a total of 1,500 miles (2,400 kilometers) of ancient roadways once radiated out from Chaco Canyon; more than 130 miles (210 kilometers) of these have been physically "ground truthed." From the air, the roads appear as narrow, dark lines running through the surrounding landscape. They are sometimes completely invisible at ground level because they are merely shallow, concave depressions only a couple of inches deep and 25 to 35 feet (7 to 10 meters) wide. They often turn with sudden, angular, dogleg jogs and are occasionally edged by low rock berms. The roads are littered with potsherds.

The longest and best-defined roads, probably constructed between 1075 and 1140, extend more than 50 miles (80 kilometers) outward from Chaco Canyon. In places the Chacoans constructed causeways; elsewhere they cut stairways into sheer cliffs. The generally straight bearings suggest that the roads were laid out—engineered—before construction, although archaeologists still debate exactly how this was done.

The roads tied far-flung regions together, moving the goods and people required to build and maintain extensive public works. Officials and bureaucrats could travel along the public roads to inspect, coordinate, and supervise. The roads themselves may have become symbols of authority, linear "banners" proclaiming affinity and cooperation, signifying participation in a system whose importance exceeded the mere sum of its parts. Apparently independent of the road system, several mesa-top signal stations found throughout the

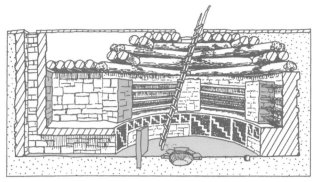

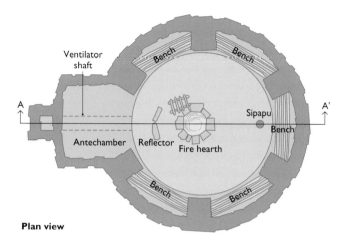

Schematic diagram of a typical Ancestral Pueblo kiva. In Pueblo religion, the *sipapu* is considered to symbolize the place of emergence from the underworld.

San Juan Basin may have provided for line-of-sight communication, presumably by smoke, fire, or reflected light.

The Kivas of Chaco Canyon

Agriculture fueled the Ancestral Pueblo lifestyle, and summer rain was necessary for the crops to grow. Anasazi people expected both rainfall and harvests in response to their prayers and solicitations. In the privacy of the kiva, people sang, prayed, and prepared for more public participation. Such ceremonies could last more than a week. Ritual derived its power from its secrecy. Gradually, from initial induction into Pueblo secret societies to the growing obligations of later adult life, the mysteries of belief were revealed.

Kivas reflect the Pueblo belief that people emerged from a previous world into this one; this process was symbolically reflected as they came up from the kiva into full view of the plaza. The kiva is an earthly representation of the original, primordial homeland, built in darkness. Into this, the ultimate cave, Ancestral Pueblo people descended through the smokehole by ladder. Set into the ground was the round, shallow, navel-like *sipapu*, symbolic of the place where the Corn Mothers emerged from the earth and ensuring spiritual access to still another world deep below.

Kivas are an omnipresent part of the contemporary Puebloan world. But the Great Kivas of the Chaco towns and their outliers were special. Many—but not all—of the Great Houses had associated Great Kivas. Several others are found up and down the canyon. The Great Kiva at Chetro Ketl, not far from Pueblo Bonito, is more than 50 feet (15 meters) in diameter. Its great curving walls held special niches, filled with strings of stone and shell beads and then sealed with masonry. There is an encircling bench, the central raised square firebox, the paired rectangular masonry "vaults," and the stair entryway. Massive sandstone disks supported equally huge roof-support timbers, each carried by hand from the distant mountains.

These Great Kivas may have involved large social units and nearby communities. At Pueblo Bonito, the two Great Kivas may each have served half of the entire pueblo. Centrally placed for all to see, the greatest of the Great Kivas extended the power of ritual beyond the human scale to the natural world.

Aerial photograph of Chacoan Great Kiva

The Chaco Phenomenon

By 900 the Ancestral Pueblo people of northwestern New Mexico had generated a sustained burst of cultural energy. Although widespread at first, this "Big Idea" soon came to be centered on the Great Houses of Chaco Canyon and is today known as the Chaco Phenomenon. By 1050 the Ancestral Pueblo had created in Chaco Canyon a place that continues to amaze: large planned towns next to haphazard villages, extensive roadways built by people who relied on neither wheeled vehicles nor draft animals. Although affluent enough to import thousands of luxury items, the Chaco Anasazi simply packed up and moved elsewhere about 1150.

Population estimates for Chaco Canyon vary widely, but a figure between two thousand and three thousand seems reasonable, based on room counts and extrapolation from modern Pueblo communities. Such a large population would have consumed a huge amount of food, and the Chaco environment is not lush. Summers are short and hot; winters are long and bitter. There is today little rainfall and hardly any drinking water. The agricultural potential is quite limited, incapable of supporting a large human population.

Most archaeologists think that the Chaco regional system may have originated to solve economic problems, possibly a food shortage triggered by environmental factors. Archaeologist James Judge suggests that this response was primarily extensive—emphasizing the drive to put more and more land under cultivation, for instance. Others see it as intensive, pointing to the development of communal ways for controlling water and other resources.

Judge has emphasized the importance of esoteric ritual knowledge as a way for the central Chacoans to integrate and control the system. Perhaps, he suggests, towns maintained their control by keeping the regional ritual calendar; this may also explain the visual communication system that seems to have

(below, and facing page)
These miscellaneous Pueblo
Bonito artifacts provide
a record of a short-lived
culture.

operated in conjunction with the road system. As this ritual hierarchy took over, social distinctions arose to separate the people living in Great Houses from the commoners who lived in scattered villages.

By 1100 a dozen large and formal Chaco towns had sprung up, and room-by-room analysis suggests that the Great Houses could have provided ample shelter for thousands of people. But today, most archaeologists believe that the Chaco buildings probably functioned mostly as public architecture, perhaps inhabited only seasonally, such as during the best growing seasons (late spring and summer).

This perspective suggests that the human population of Chaco Canyon fluctuated dramatically. Judge suggests that floods of relatives showed up in Chaco Canyon during what he calls "pilgrimage fairs," when Chacoan people from the hinterlands would visit the canyon, which had become increasingly important in their ritual and ceremonial lives. In good years, travelers may have brought extra food on the pilgrimage, perhaps in exchange for esoteric ritual knowledge.

Lynne Sebastian offers an alternative model to explain the emergence of Chacoan complexity. She believes that other reconstructions have relied too heavily on analogies to modern Pueblo people. Even if historic Pueblo society completely lacked social differentiation and political hierarchy—a debatable proposition—this need not preclude such differentiation and hierarchy in twelfth-century Chaco Canyon. She also takes exception to redistribution-based explanations for the Chaco system, arguing that, lacking draft animals, redistributing people is more efficient than redistributing bulk subsistence items.

Sebastian begins her explanation of Chacoan complexity with the extremely fine-grained rainfall reconstruction created by the University of Arizona's Tree-Ring Laboratory. Sebastian believes that the transition from local to regional leadership took place during times of significantly increased rainfall in the ninth century. Leadership was situational, based mostly within traditional kinship structures. But as ecological conditions improved, some groups found themselves producing crop surpluses. Those farming the best land were suddenly in a position to capitalize on their success. These opportunistic local leaders used their agricultural good fortune to cement alliances with those less fortunate.

By the eleventh century, a powerful elite had emerged in Chaco Canyon. Their power was both reinforced and expanded by continued competition for the loyalty of the less affluent. In time this competition escalated beyond simply hosting the feasts and festivals. According to Sebastian's interpretation, the monumental Chacoan Great Houses, including Pueblo Bonito, were erected not to store surpluses or house visitors, but to emphasize the sociopolitical power of those who built them. The massively scaled Great Houses and Great Kivas, the arrow-straight roads and earthworks were all flamboyant expressions of power built by competitive patron groups.

138
—

CHAPTER ELEVEN

Chaco Abandonment

Regardless of why the Chaco Phenomenon arose, it did not last long. Chaco Canyon was largely abandoned by 1140—at least the tree-ring dates suggest that no new construction took place after this date. If people kept living in the canyon after this date, there is little evidence for it. Perhaps the Chaco Anasazi reorganized socially. Or maybe they intensified economically. Or maybe they just moved out, shifting northward to the San Juan River. Whatever happened, by the mid-twelfth century the central Chaco system fell apart. But the hinterland was doing just fine.

About 1240 there was a renewed intensity of construction in Chaco Canyon. Both the architectural and ceramic styles come to resemble those of the northern (Mesa Verde) area. Some believe that whereas everyday life in Chaco carried on, the regional focus shifted toward the San Juan Basin to the north. Clearly, in the 1200s Chaco was neither the cosmological nor the economic center of the Ancestral Pueblo world.

139
—

Society and agriculture were severely disrupted in about 1300 by a series of major droughts across the Chacoan landscape. For five decades, life-giving summer rains stayed away. As the Chaco world dried up, people naturally intensified their efforts to extract yet more from the earth and sky. More and more Ancestral Puebloans were drawn into the web of concern. Soon the area became depopulated, the Ancestral Puebloans probably moving eastward. Navajo people moved in during the 1700s and stayed until 1947, when Chaco Canyon National Monument was fenced by the Park Service.

For decades, the abandonment of Chaco Canyon has been seen in largely negative terms: Ancestral Pueblo people were "pushed out" because of a deteriorating environment, collapsing economic systems, and social chaos. But as we saw in Chapter 10, archaeologists William Lipe, Bruce Bradley, and others are rephrasing the question, going beyond the "pushing out" to look at social forces that may have "pulled" the Ancestral Pueblo away from Chaco.

Whatever the reason for their departure, the Chaco people did not die out. There was no violence. They evacuated their towns in orderly fashion, taking the most useful and portable material goods with them, and they reorganized, shifting their social and religious priorities to meet the challenges of survival. In their new tribal territory, Anasazi descendants would encounter sixteenth-century Spanish explorers who reported that 50,000 Pueblo people lived in more than a hundred towns along the margins of New Mexico's San Juan Basin and the Rio Grande drainage.

Further Reading

There is a huge body of published literature on Pueblo Bonito and the archaeology of Chaco Canyon. I particularly recommend *Chaco Canyon: A Center and Its World* by Mary Peck, with essays by John R. Stein, Stephen H. Lekson, and Simon J. Ortiz

(Santa Fe: Museum of New Mexico Press, 1994); *Chaco & Hohokam: Prehistoric Regional Systems in the American Southwest*, edited by Patricia L. Crown and W. James Judge (Santa Fe, NM: School of American Research, 1991); *Roads to Center Place, A Cultural Atlas of Chaco Canyon and the Anasazi*, by Kathryn Gabriel (Boulder, CO: Johnson Books, 1991); and *In Search of the Old Ones: Exploring the Anasazi World of the Southwest* by David Roberts (New York: Simon and Schuster, 1995).

Other useful discussions are *Chaco Canyon: Archaeology and Archaeologists* by Robert H. Lister and Florence C. Lister (Albuquerque: University of New Mexico Press, 1981); *The Chaco Anasazi: Sociopolitical Evolution in the Prehistoric Southwest* by Lynne Sebastian (Cambridge: Cambridge University Press, 1992); and *The Chacoan Prehistory of the San Juan Basin* by R. Gwinn Vivian (New York: Academic Press, 1990).

Linda S. Cordell provides a more general context in *Ancient Pueblo Peoples* (Montreal and Washington, DC: St. Remy Press and Smithsonian Books, 1994).

140
—

Further Visiting

Chaco Culture National Historical Park (northwestern NM) contains the ruins of Pueblo Bonito as well as several other Great Houses and thousands of smaller sites. The park can be reached from the town of Thoreau (about 60 miles [100 kilometers] north) or from Gallup to the south. Hiking trails lead into the nearby backcountry.

Pueblo Grande

A.D. 500–1450

Hohokam culture

in Phoenix, Arizona

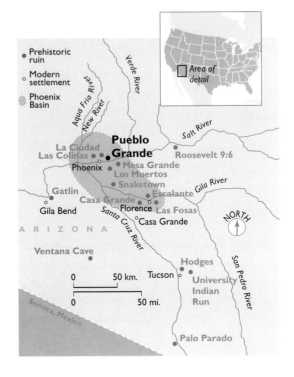

Pueblo Grande, located in Phoenix, Arizona, was one of the largest and most important villages in the Salt River Valley for more than seven centuries. By A.D. 1350 it was the preeminent Hohokam community north of the Salt River. Covering more than 1 square mile (2.5 square kilometers), Pueblo Grande had dozens of major domestic and ceremonial buildings, at least two ballcourts, and a three-story adobe tower now called the Big House. At the southern end was a massive stone and adobe platform mound. One of the largest monuments ever erected by the Hohokam, the mound stood more than 20 feet (6 meters) high; its base would cover a modern football field.

Pueblo Grande also looms large in O'odham (Pima) Indian tradition. As they came from the east, the O'odham ancestors attacked the arrogant Hohokam rulers, whom they called Sivanyi. Elder Brother, a legendary O'odham figure, had been insulted by the Sivanyi, who bragged of their knowledge and plotted to assassinate Elder Brother. In support of Elder Brother, the Akimel and Tohono O'odham (Pima and Papago) people attempted to destroy all the great Hohokam villages of the Salt and Gila River valleys.

After demolishing a number of lesser Hohokam communities, Elder Brother's forces came to Pueblo Grande, where the Sivanyi put up a bold resistance. After a pitched battle, the Sivanyi leader known as

Aerial photograph (1928) of the Pueblo Grande platform mound

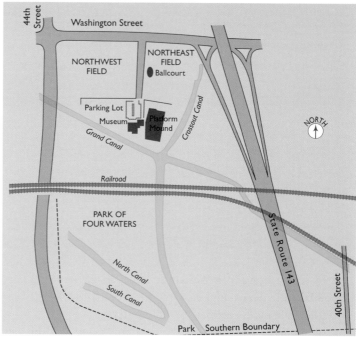

Site map of Pueblo Grande

Yellow Buzzard was slain, along with his Hohokam followers. The legendary battle of Pueblo Grande is significant because it marks the last major resistance to the O'odham takeover of the Salt River Valley.

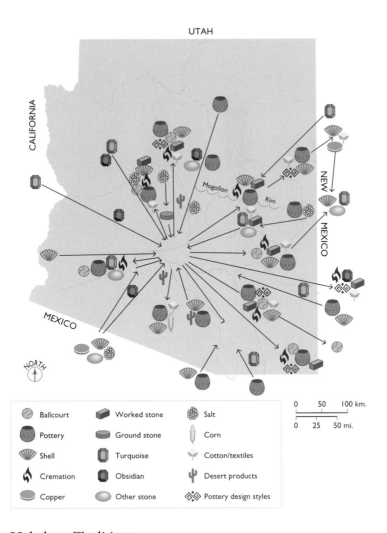

UTAH

CALIFORNIA

NEW MEXICO

Mogollon Rim

MEXICO

NORTH

🏐 Ballcourt	🧱 Worked stone	❋ Salt
🏺 Pottery	⬭ Ground stone	❘ Corn
🐚 Shell	◆ Turquoise	❦ Cotton/textiles
♨ Cremation	● Obsidian	🌵 Desert products
⬭ Copper	◯ Other stone	◈◈ Pottery design styles

0   50   100 km.
0   25   50 mi.

The Hohokam Tradition

The Hohokam people lived in the blistering deserts of southern Arizona and northern Sonora. The name "Hohokam" is an O'odham word literally translated as "those that are gone," but usually is given the more generic translation of "ancient" or "old ones." Hohokam communities straddled major continental trade routes—from the California coast to the Great Plains, from the high civilizations in Mexico to the resource-rich Rocky Mountains—and engaged in far-flung, if rather small-scale, barter for merchandise as diverse as buffalo and deer skins, sea shells, turquoise, obsidian, rare minerals, finished textiles, salt, exotic feathers, and ceramics.

After looking at the precise year-by-year chronology available for much of the Anasazi area, southwestern archaeologists have become a bit spoiled, at least in terms of temporal controls. In the Hohokam area specialists bemoan the second-rate (and still controversial) chronological controls available for Hohokam sites. To some extent, their complaints seem justified.

This Hohokam red-on-buff jar is one of many examples of the fine ceramic work at Pueblo Grande.

PUEBLO GRANDE

ON HOHOKAM ORIGINS

For years archaeologists believed that Hohokam origins went back to the late Archaic hunter-gatherer lifeway. The hunter-gatherers gradually came to rely on part-time maize agriculture to supplement their traditional diet of game, piñon nuts, cactus fruit, and hard-shelled seeds. In this view, the agricultural crops were introduced as backups, buffers against failure of the wild staple foodstuffs. The transition to sedentary agricultural villages spanned centuries, if not millennia.

Recent research along the floodplain of Tucson's Santa Cruz River has changed this view. An important series of archaeological projects, conducted within the right-of-way of the Interstate 10 corridor through Tucson, has disclosed several deeply buried sites that have forced archaeologists to reconsider the transition to farming-based village life. Dating to between 1200 B.C. and A.D. 150, these so-called "Cienega phase" sites suggest that maize, beans, and squash (and possibly tobacco and cotton) may have arrived during a large in-migration of agricultural people.

So dramatic are these new finds that archaeologist Bruce Huckell suggests that the term late Archaic be scrapped in favor of an early Agricultural Period, highlighting the significance of these huge semipermanent villages. Over four years, the Center for Desert Archaeology completed seven digs, including important excavations at the massive Santa Cruz Bend site. Directed by Jonathan Mabry, the archaeological teams found nearly two hundred pit structures, including a "Big House" (28 feet [8.5 meters] in diameter) and associated plaza; these public areas may have been used for religious rites and political gatherings, serving as the central focus of the 300 B.C. settlement. Extended families lived in circular house groups, sharing courtyards and storehouses. The sheer size of these settlements has overturned our views of pre-Hohokam society.

The Santa Cruz Bend villagers were manufacturing pottery as early as 800 B.C., at least a millennium earlier than previously recognized. Their water-control ditches presage the well-known Hohokam irrigation systems. Their elaborate jewelry demonstrates a surprising degree of long-distance exchange. A tobacco-filled stone pipe (dating about 350 B.C.) is the earliest evidence of tobacco use in North America.

Clearly, the Cienega phase people of southern Arizona had developed a settled village lifestyle that predated that of the better-known Hohokam and Anasazi cultures. Although Mabry thinks that these early villagers might be among the ancestors of later Hohokam people, the picture is as yet far from clear.

Chronological problems have dominated Hohokam archaeology for years. Because desert woods are unsuitable for tree-ring dating, Hohokam chronology has been built from a host of less precise techniques, including radiocarbon and paleomagnetic dating, augmented everywhere by ceramic stylistic dating methods. Still, the Hohokam sequence can be divided up into two hundred-year periods, which is superior to what is available to archaeologists working in many parts of North America. It is only by comparison with the various

Anasazi microchronologies that the Hohokam sequence suffers. By general North American standards, the Hohokam archaeologists still have it pretty good.

Only in the final decades of the twentieth century have archaeologists transcended chronological issues to look more deeply into the meaning of Hohokam behavior. Excellent data are now available to document patterns of Hohokam demography and intracommunity patterning, and Hohokam archaeologists are sketching out a picture that is nothing short of astounding: The complexity seen in these parched desert sites clearly rivals—and at times uncannily mirrors—contemporary Anasazi developments, particularly those in Chaco Canyon.

But we are getting ahead of the story. New research suggests that Hohokam roots run back into the so-called early Agricultural period of southern Arizona. The earliest distinctive Hohokam pattern, however, seems to have been established by about A.D. 1. Expanded farming productivity derived in part from successful experiments with innovative diversion dams, ditches, and levees. These earliest Hohokam people lived in pit houses, some of which assumed hefty proportions—up to 150 square feet (13 square meters). By 700, the end of the so-called Pioneer period of the Hohokam, farming communities were prospering along the major river systems of the Phoenix Basin, and the Hohokam heartland covered roughly 4,000 square miles (10,360 square kilometers). Communities probably ranged in size from a few extended families to more than several hundred people. Trade networks exported worked stone (particularly obsidian), salt, pigments, turquoise, and other desert resources as far as central Mexico. From the southland, they received stone mosaic mirrors, marine shells, birds with flamboyant plumage, and carefully crafted copper bells. They also traded with their Anasazi contemporaries.

A new feature of Hohokam architecture, distinctive large earthen mounds, appeared sometime before 800. These slope-sided accumulations of trash and desert soil with plastered tops may once have served as dance platforms. Hohokam of this period also continued a long-lasting trend by cremating their dead.

Between 700 and 1100 villages became larger. There was more intensive canal irrigation, the population grew, and ceremonialism became considerably more elaborate. Trade networks expanded far beyond previous frontiers, funneling scarce and expensive resources to talented Hohokam artisans. Artistic excellence was expressed in more diverse media: stone, bone, shell, and ceramics. The largest Hohokam villages during this period exceeded 500 acres (200 hectares) in size, hosting perhaps a thousand people. The villages, each with their own ballcourts, were spaced evenly at about 3-mile (5-kilometer) intervals along the main irrigation canals.

At this point, understanding Hohokam archaeology depends upon deciphering the meaning and distribution of two key facilities in Hohokam society: irrigation networks and ballcourts. Consider first the waterworks.

Some archaeologists believe that Hohokam cremation rituals involved melting lead silicates on stone palettes such as this one.

145
—

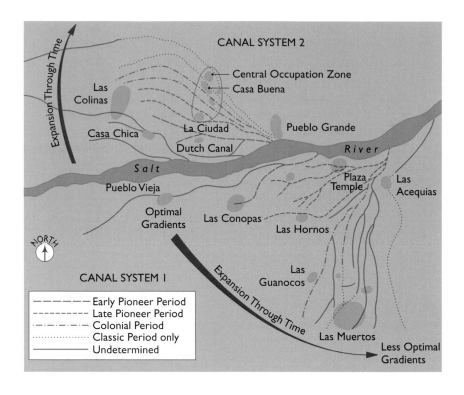

Hohokam Waterworks

Impressive and increasingly advanced technology harnessed the most scarce
Hohokam resource of all: water. Their ability to capture and harness water-
power through canal irrigation was unrivaled in Native North America, and
it allowed them to move their agriculture from the floodplains, where nature
provided water, to the terraces, where considerable human ingenuity was

Figurines such as these
Hohokam desert bighorn
effigies are often found in ritual caches.

required. More than 900 miles (1,450 kilometers) of canals have been mapped in the Salt River Valley alone; they date from about A.D. 1 to 1500. Other systems are known from the Gila River Valley, and a smaller number have been found in the Tucson Basin as well.

By A.D. 500–600, the Salt River Valley canal system had reached a massive scale. Some canals were huge even by today's standards, measuring up to 75 feet (22 meters) across at the top. Many individual canals reached 7 miles (11 kilometers) in length, and a few extended more than 20 miles (32 kilometers). Hundreds of smaller ditches snaked across the Sonoran desert, further expanding the horizons of the Hohokam agricultural base. The modern city of Phoenix employs a canal system virtually superimposed on the early Hohokam plan for diverting water from the Salt River, a mute and unintentional tribute to the Native American engineers who came before.

Canal irrigation required not only a tremendous initial investment in backbreaking labor but constant monitoring and maintenance. People had to open and close the floodgates; downstream, they had to muck out the canals periodically; and they had to repair flood damage rapidly to prevent the summer and fall crops from withering. Elaborate social mechanisms must have arisen to provide the necessary leadership and coordination to meet the demands of effective water management in the parched desert setting. Particularly challenging must have been the redistribution of agricultural products among villages of vastly unequal rank.

Paleoenvironmental reconstructions suggest that the optimal conditions for establishing canal irrigation prevailed from 900 to 1051 and that a series of flood and drought episodes made canal irrigation much less effective between 1197 and 1355. As you will see later in this chapter, some investigators feel that problems with these irrigation systems may have contributed to the ultimate downfall of the Hohokam.

The Ballcourts

Hohokam ballcourts appeared about 800 or so and proliferated until about 1150, after which no more ballcourts were built. But during their heyday, the distribution of ballcourts defined, in a real sense, the boundaries of the Hohokam regional system. Adobe embankments surrounding the largest ballcourt—at Snaketown, Arizona—stood 16 feet (5 meters) high and stretched nearly 200 feet (60 meters) on a side. At least five hundred people could have stood on these elevated sidelines to watch the goings-on. Two smaller ballcourts can be seen today at Pueblo Grande.

There is no ethnographic data on the use of Hohokam ballcourts, so it is difficult to deduce exactly what activities and events specifically took place in them. A stone ball about the size of a baseball was found in the north ballcourt at Pueblo Grande, and stone markers have been placed along the floor, possibly lining out zones of play. The Hohokam ballgame may have begun as

A large proportion of Hohokam human effigies represent females, suggesting to some archaeologists a connection to fertility and agricultural productivity.

an imitation of the ritual athletics in central Mexico, where Hohokam people perhaps observed the proceedings firsthand. But once back home, the Hohokam seem to have developed their own version, including the unique construction of their ballyards.

While it seems likely that both intramural and intervillage contests took place, we do not know the rules—it was probably a stick or kicking game—or the social mechanisms that tied Hohokam populations living in communities without ballcourts to communities that did. Still, at one level at least, ballcourts provided a socially sanctioned mechanism for legitimizing face-to-face interaction between distant yet somehow related Hohokam populations.

Many archaeologists assume that the ballgames served some sort of integrative function because there are no architectural barriers or restrictions to keep people from seeing them. A few of the northern ballcourts also seem to have been built by other cultures, perhaps to integrate themselves into the Hohokam regional system. By 1050 the ballcourt network contained more than two hundred venues, all of which appear to have been somehow connected to trade.

Whatever the specifics, it seems that the ballgame transcended the individual community, political alliances, or economic cooperation. David Wilcox believes that the Hohokam ballcourts are best seen as public architecture, in which ritualized events took place, probably also involving ceremonial exchanges. While the ballcourts surely could have served as arenas for ballgames, they were also symbolic spaces, excavated into the earth like Pueblo sipapu, perhaps as portals to the underworld. These ballcourts may have facilitated ceremonial exchanges among very different "ethnic" groups, without the need for longer-term political or economic dependencies. Ballcourt ritual probably served the latent but critical function of facilitating greater dissemination of goods, services, and presumably marriage partners.

Beyond the community level, the units of Hohokam society were held together by two overarching partnerships. One was the ballcourt organizations. The other was the social controls that maintained the irrigation necessary for basic Hohokam survival.

Hohokam Social Organization

For more than a century, archaeologists have tried to make sense of Hohokam social organization. Dozens of scholars have looked at architectural data, settlement pattern evidence, artifacts, and mortuary patterning. In the 1890s, Frank Hamilton Cushing interpreted Classic period Hohokam social organization as nonegalitarian (that is, ranked). Then, during the 1930s, archaeologists Harold Gladwin and Emil Haury developed a model of Hohokam society as basically egalitarian, much like ethnohistoric Tohono O'odham communities that lived in dispersed autonomous communities along major river margins.

In the closing decades of the twentieth century, the tempo of Hohokam archaeology has accelerated dramatically. Several explanations of Hohokam social organization now exist. Although some archaeologists still view the Hohokam in largely egalitarian terms, many others argue for greater social complexity, suggesting that Hohokam society must have been ranked, perhaps operating at the chiefdom or even state level. Others believe that no evidence exists for a single chiefdom or hierarchy; perhaps the Hohokam consisted of several cooperative and/or competitive irrigation communities, ideologically unified in some fashion.

Some prefer a more theocratic model of sociopolitical organization, in which full-time priests dramatically plied their trade atop the prominent platform mounds. Perhaps these priests had a hand in controlling the massive irrigation systems established along the Salt River Valley. The extensive mortuary data from Pueblo Grande—more than one thousand individuals—shows no significant social differentiation and relatively little acquisition of personal wealth. To be sure, some of the Hohokam men, women, and children were occasionally interred with valuables, but nothing to compare with, say, the Ohio Hopewell (Chapter 7). Many today consider such "striking" burials to be those of shamans or priests.

The Hohokam Demise

A shrinkage of the Hohokam regional systems after about 1150 corresponds closely to the collapse of the ballcourt system, which died out and was abandoned by 1200 or so. Despite many parallels to contemporaneous Chaco Canyon, what happened next is uniquely Hohokam. Rather than abandoning its core, Hohokam society may have reorganized trade networks and political alliances but left the bulk of the population in place.

During the Classic Hohokam period, between about 1100 and 1450, some traditions fell into decline. A few centers of Hohokam life were abandoned or shifted location (particularly by 1050 or so), craftsmanship trailed off, and few new ballcourts were built. Although cremation continued throughout the Classic period, perhaps two-thirds of the population practiced new burial forms, primary inhumation. Curiously, however, moundbuilding surged during this period, with more and larger mounds being built. More than forty mound sites have been detected in Hohokam territory, the largest of which, at Pueblo Grande, contains up to 32,000 cubic yards (24,000 cubic meters) of fill.

Ceremonialism during Classic Hohokam times became less public. Compounds, mounds, and walled plazas suggest a heightened sense of secrecy and seclusion. People built elaborate mound-top houses and adorned themselves with exotic turquoise, worked marine shells, and gaudy feathers. Some see such elaborate dwellings and ornamentation as concrete signs of elite privilege. But others believe that all Hohokam people wore shell jewelry and other items

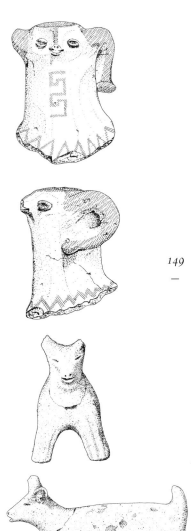

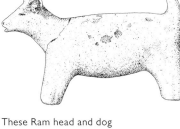

These Ram head and dog effigies are from Pueblo Grande. The ram's head was formed on the rock of a red-on-buff jar. Traces of red pigment occur along the horns and eyes of the ram.

of personal ornamentation. The so-called "elite dwellings" may instead have been special-purpose rooms used, perhaps, as priests' living quarters or for storage of ceremonial objects too powerful to be left untended.

Disaster must have struck the Hohokam heartland about 1400. Around that time the people abandoned once-flourishing towns. Why? Nobody is sure. The paleoenvironmental evidence of increased flooding and intervening dry years in the mid- to late 1300s may offer a clue. Alternating droughts and floods would have had disastrous effects on the canal system. Perhaps there was an increasing salt buildup in the soils as well. Recent bioarchaeological analysis of human remains from Pueblo Grande suggests that the Hohokam people were suffering from extreme nutritional stress and were unable to self-sustain a population at reproductive levels. For the Pueblo Grande population as a whole, the mean life expectancy was slightly over fifteen years of age; only half of the children born reached their first birthday. It has become increasingly clear that a general decline in the health and nutrition of Hohokam people exacerbated the effects of both social and natural impacts to their society during the late fourteenth and early fifteenth centuries.

Whatever the causes, the demographic consequences were dramatic. Perhaps five thousand O'odham people lived in the Phoenix basin during the seventeenth century. But estimates of the population just two centuries earlier range from 24,000 to more than 50,000 people. By the time Jesuit Father Kino said mass at Casa Grande in 1694, this great village already lay in ruins.

There is no consensus about whether the Hohokam were ancestors of the O'odham Indians who greeted the first European explorers in the Sonoran Desert. The Hohokam and the historic-period O'odham built similar dwellings and council houses, participated in ballgames, had similar subsistence bases, and buried their dead in similar ways. For years Emil Haury, the dean of Hohokam archaeology, argued that the O'odham people descended directly from the Hohokam. But other readings of oral history provide a different answer, questioning the so-called O'odham-Hohokam continuum. Yet new ceramic-based evidence supports the continuum interpretation.

Whatever caused this population to decline so radically in post-Hohokam times, it is clear that the decline drastically simplified the Hohokam lifestyle—perhaps directly into the historic O'odham existence. Another argument holds that the Hohokam were a multi-ethnic community that could have left multiple descendants. At the strictly archaeological level, the question remains open.

But as we saw at the beginning of this chapter, the O'odham people see their own history somewhat differently. The official O'odham (Pima) position argues for the O'odham-Hohokam continuum. O'odham oral tradition holds that the Hohokam leaders were evil, and that their own ancestors arrived from the east, destroying the platform mounds and the leadership hierarchy that built them. In this view, the destruction of Pueblo Grande is seen as civil strife within a long-term cultural tradition.

Further Reading

The best general overview is *Desert Farmers at the River's Edge: The Hohokam and Pueblo Grande* by John P. Andrews and Todd W. Bostwick (Phoenix: Pueblo Grande Museum and Cultural Park, 1997). I also recommend *The Hohokam: Ancient People of the Desert* by David G. Noble (Santa Fe: School of American Research Press, 1991); *Chaco & Hohokam: Prehistoric Regional Systems in the American Southwest*, edited by Patricia L. Crown and W. James Judge (Santa Fe: School of American Research, 1991); *Exploring the Hohokam: Prehistoric Desert Peoples of the American Southwest*, edited by George J. Gumerman (Albuquerque: University of New Mexico Press, 1995); and *The Hohokam: Desert Farmers and Craftsmen* by Emil W. Haury (Tucson: University of Arizona Press, 1976).

Further Visiting

The Pueblo Grande Museum (downtown Phoenix, AZ) contains the large rectangular platform mound of caliche-adobe. Two or more ballcourts also stood here, as did a major multistory building.

Cahokia

A.D. 800–1350

Mississippian culture

in East St. Louis, Illinois

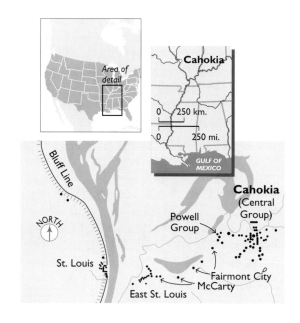

Cahokia was the largest pre-Columbian town in North America—five times the size of its nearest competition. The site covers 5 square miles (13 square kilometers) of the rich floodplain known as the American Bottom, where the Mississippi and Missouri rivers joined forces near present-day St. Louis. This huge complex of Mississippian culture was surrounded by fertile soils and plentiful wildlife. There is ample evidence of large-scale construction projects, major residential areas, open plazas, protective palisade walls, elite burials, and exotic artifacts. But we do not know what Cahokia's inhabitants called it. The word "Cahokia" comes from a subtribe of the Illini Indians who apparently came on the scene after the demise of Cahokian culture.

Between A.D. 800 and 1350 a number of competing chiefdoms existed in the American Bottom, sometimes consolidating into a single paramount chiefdom, at other times splintering and warring with one another. At its peak, Cahokia was home to 10,000 to 15,000 people, and perhaps tens of thousands more lived on the surrounding floodplain.

Investigators have viewed the emergence of Mississippian culture at Cahokia from many perspectives. Some have seen it as a case of *in-situ* evolutionary development: an orderly emergence from late

Woodland populations, with minimal outside contact and stimulus. Others believed that the development must have been stimulated by migration and contact from populations to the south. Still others argued the developments in the American Bottom reflected broader cultural processes taking place throughout much of the Mississippi River Valley, with Cahokia emerging perhaps as a market and center for redistribution. These three models employed the notion of regional interactions, not unlike the Hopewell Interaction model discussed in Chapter 7.

But recent research has rejected the possibility of any large-scale population movements from the outside to explain Mississippian origins. A new consensus has emerged, articulated by George Milner, Thomas Emerson, Timothy Pauketat, and others. These archaeologists view Cahokia dominance as much briefer than previously thought. Cahokia's political power came from negotiated alliances between economically self-sufficient "rural" populations that lived in surrounding town-and-mound centers. To a large degree, the Cahokian rural populations were independent of one another and of the elite living at Cahokia proper. But, Pauketat and Emerson emphasize, during the ascendency of Cahokia, these rural populations were tightly controlled. This chapter explores some aspects of this new model.

The Emergent Mississippian Period (A.D. 800–1050)

By about A.D. 700 the indigenous late Woodland people had established compact villages in the Cahokia area. During the intensive transformation of social, political, religious, and economic organization of the next three centuries, the people living in the American Bottom would adopt full-scale maize horticulture, establish nucleated villages, and develop many of the material traits that would come to define the subsequent middle Mississippian culture.

Between 800 and 1050, the emerging Mississippian economy transcended the traditional Woodland-era cultivation of native plants to focus on maize. Soon maize-based farming spread over eastern North America and helped support more complex sociopolitical structures. Ultimately, maize horticulture would support the evolving Iroquoian confederacy of the Northeast, the Fort Ancient polities along the middle Ohio River Valley, and the diverse array of Mississippian chiefdoms that controlled the river valleys of the Southeast and Midwest. After European contact, maize sustained the Creek and Choctaw tribes of the Southeast and the Mandan and Arikara of the Plains. From the time it arrived from Mexico, maize dominated southwestern diets. But in the eastern Woodlands, more than six centuries elapsed between its introduction and its dominance. Archaeologists are still attempting to explain this lag. And the role of maize in Cahokia's development remains controversial. Emerson and Pauketat argue strongly that sociopolitical rather than agricultural factors spurred the site's emergence around 1050.

The adoption of maize-based agriculture at places like Cahokia was a

154
—

watershed event in American history. Unlike their European counterparts, Mississippian people did not domesticate draft animals. Instead, Native American farmers tilled their land by hand and hoe, in fields usually located along fertile river valleys or abandoned levee meanders. Cahokia was deliberately sited near prime farmland and connected to the water and land routes tying the city to neighboring and distant settlements.

The so-called Emergent Mississippian period was a time of extreme cultural variation. The half-dozen or so civic and ceremonial centers of the American Bottom competed at both local and regional levels. By the early eleventh century, Cahokia emerged as the paramount civic-ceremonial center in the area. The next century was politically unstable, as rival elite factions vied for power and prestige.

Between 925 and 1050, both Cahokia and the rural landscape were dominated by kin-based societies. Settlements were commonly organized into courtyard clusters of small houses of single-post construction. Communities varied from small, isolated farmsteads to much larger settlements. At Cahokia proper, variability can be seen between communities, suggesting that some residents abruptly moved away from farming and hunting toward dependence on external supplies and/or tribute. These communities were probably simple chiefdoms and villages that gradually grew larger.

Archaeologists use the term "Mississippian" for the hundreds of farming societies that thrived between about A.D. 800 and 1500 throughout the Tennessee, Cumberland, and Mississippi River valleys. Traditionally scholars recognized Mississippian culture by certain distinctive features: characteristic pottery, usually tempered with crushed mussel shell; village-based maize horticulture; construction of large flat-topped mounds, commonly situated near the town plaza; and stratified social organization embodying permanent and probably hereditary offices. Mississippian people also adopted the bow and arrow, explicitly connected their religion to agricultural productivity, often worshiped a fire-sun deity, and engaged in intensive long-distance exchange.

More recently, however, archaeologist James B. Griffin suggested that Mississippian societies should be perceived more broadly, with special attention to the following tendencies: active involvement in numerous cultural innovations between A.D. 700 and 900; incorporation of these innovations through contacts with other groups; construction of planned permanent ceremonial centers, towns, and hinterlands; a hierarchical social, political, and religious system; a religious worldview that emphasized interaction with a spirit world, as expressed through a rich iconographic tradition involving artifacts of marine shell, copper, ceramics, and stone; an extensive trade network; and achievement of a cultural climax or "crest" between 1200 and 1500.

However we define Mississippian, we know that during their heyday the Mississippian elites presided over breathtaking ceremonial centers, places today called Cahokia, Moundville, and Spiro. The Mississippian aristocracy was invested with power by the thousands upon thousands of farming people who lived in smaller palisaded hamlets and farmsteads. Although much of eastern North America did not participate in full-blown Mississippian culture, the entire region was to some extent dependent upon Mississippian-style economics. Descendants of the great American Indian confederacies of the American southland, including the so-called Five Civilized Tribes, are deeply rooted in their Mississippian ancestry.

155
—

The Middle Mississippian Period (1050–1300)

By the end of the Emergent Mississippian period, rural populations were moving away from the countryside, presumably relocating to larger nucleated centers. A select few of these clustered villages abruptly became major centers of both power and population.

During the middle Mississippian period the American Bottom landscape was transformed by the appearance of a centralized elite power base at Cahokia, which became increasingly urbanized and nucleated, almost certainly because rural populations relocated into the central core area. These formerly peripheral subgroups, now living at Cahokia, provided the elite with direct control over the rural farmers, who were now part of the Cahokia power structure. The paramount elite at Cahokia had created a political and religious network to ensure a steady food supply.

Quite rapidly, the nearest town sites shrank and larger temple towns arose in the American Bottom some distance from Cahokia. Most of the competing chiefdoms were absorbed, and the earlier household courtyard communities broken up and replaced by small, dispersed farmsteads and households.

These small agricultural settlements maintained gardens and larger agricultural fields.

The paramount elite also superimposed a series of "nodal centers" across the rural landscape. Special-purpose sites and large nucleated centers with both civic and ceremonial duties created a functional political fabric with a degree

WOODHENGES: THE SUN CIRCLES

A remarkable triumph of Native American science and engineering was accidentally discovered not far from Monks Mound in the early 1960s. Knowing that an interstate highway was scheduled to run through the heart of Cahokia, a team of professional archaeologists worked feverishly to keep ahead of the bulldozers. Warren Wittry noticed several large oval-shaped pits seemingly arranged in arcs. Suppose, he thought, that each pit had held an upright post. If so, then perhaps the Mississippians at Cahokia had devised a kind of sun calendar. He called his find the Woodhenge.

Wittry predicted that—if his hunch were correct—other undiscovered pits would be found along the rest of the arc. Follow-up excavations did indeed expose a series of post pits exactly where Wittry had predicted. Some of the pits even contained remnants of the original cedar posts, stained with red paint.

Today archaeologists recognize a number of sequential "woodhenges" at Cahokia. Their diameters range from 240 to 450 feet (75 to 140 meters). Each has a center post and up to forty-eight posts defining its precisely circular perimeter. These telephone pole-sized posts were set into deeply dug pits, each the size of a modern bathtub. Wittry suggested the posts may have been 23 to 32 feet high (7 to 10 meters); if so, each would have weighed several thousand pounds. Melvin Fowler also suggested that a similar post circle existed at Mound 72.

Recent chronological studies show that the Woodhenge was built and used over, at most, a single century, between 1100 and 1200. Based on superposition, Pauketat believes that the earliest and largest of the woodhenges at Cahokia consists of sixty posts.

Interpretations of Cahokia's woodhenge structures vary considerably. Wittry argued for a calendrical function, based on tracking of solar positions; for him, these were "sun circles" writ large. Robert Hall compared the woodhenges to Plains Indian "world center shrines," symbolizing an inherent cosmic order, related to annual fertility rites and perhaps to calendrical reckoning. Others have challenged the calendrical hypothesis, suggesting instead that the woodhenges may have helped the Cahokia architects lay out the basic site plan.

If these Mississippians did follow a calendar, much of their knowledge has been lost, and today we cannot fully know how these woodhenges operated. Perhaps a sun priest stood in the center of the circle and "read" the calendar by observing where the sun rose. The simplest such calendars need only three points: one each marking the first days of winter and summer (the solstices), and one post halfway between to indicate the first days of spring and fall (the equinoxes). The most spectacular sunrises occur at the equinoxes, when the sun comes up due east. Viewed from the Woodhenge at Cahokia, these sunrises take place directly over the top of Monks Mound. What a powerful sight: the elite ruler's residence giving birth to the sun.

This earliest woodhenge at Cahokia was reconstructed in 1985 on its original location. Today you can walk among the forty-eight regularly spaced uprights, defining a circle 410 feet (125 meters) in diameter. But the remaining dozen posts are still a mystery. Maybe they marked special festival dates, probably relating to the agricultural cycle. They may also have served to enclose the area in which the sacred woodhenge ceremonies were conducted. Perhaps, as others suggest, some of the "extra posts" helped predict alignments of certain stars, or even functioned as eclipse predictors. Pauketat prefers to call the woodhenge structures "post-circle" monuments. They were almost certainly elite-controlled devices, built to emphasize the growing social distance between elite and nonelite households.

of internal cohesiveness, encouraged by kin ties and an elite that controlled supernatural and natural resources. Clearly, the elite at Cahokia was exerting coercive power over formerly scattered groups of the hinterland.

Local and exotic items of wealth became concentrated at Cahokia, where the population skyrocketed from perhaps 1,300–2,700 people during the Emergent Mississippian period to 10,000–15,000 during the middle Mississippian. Political consolidation was inherently unstable; various elite groups vied with each other to control resources and power. Between 1200 and 1250, a palisade wall was erected to protect Cahokia's central precinct. This 2-mile (3-kilometer) long wooden palisade may have reached a height of 12 to 15 feet (4 to 5 meters); if so, it would have required 15,000 to 20,000 logs, and some archaeologists have stressed the enormous environmental impact of cutting so many trees.

During the height of the middle Mississippian period, the elite appropriated civic and religious power from commoners. By the 1100s downtown Cahokia had a large sacred landscape, built by concerted public labor. In the hinterlands, the central elite's control and power were reflected in ceramic iconography, which cleverly combined elite-mediated symbolism with more traditional "folk" cosmology. Specialized religious, mortuary, and civic architectural facilities were also constructed in the countryside. The elite class would become increasingly sacred, as seen in the elaborate symbolism of Cahokia monuments, buildings, and ceramic iconography.

Monks Mound

Lay people often refer to the Mississippians as Mound Builders—an apt, if outmoded, description of what took place at Cahokia. The immediately adjacent area may have once contained 120 Mississippian mounds. A few mounds were used for burying the dead, but most were devoted to ceremonies of the living. In houses atop these temple mounds, Native American aristocrats presided over the rituals that codified the Mississippian lifeway. Many of the Cahokia earthworks have been thoughtlessly destroyed, but sixty-five still survive within the boundaries of the Cahokia Mounds State Historic Site.

Then as now, the Cahokian landscape was dominated by Monks Mound, the largest earthwork ever constructed in the Americas and perhaps the largest in the world. Monks Mound was not built by monks; the name recalls nineteenth-century Trappist monks who planted gardens and orchards on its expansive terraces.

Built in stages over three centuries, Monks Mound covers a city block (17 acres [7 hectares]), contains 22 million cubic feet (615,000 cubic meters) of earth, and at nearly 100 feet (30 meters) reaches the height of a ten-story building. Recent work has shown that major portions of the earthwork were in place by 1000–1100 and that construction was well under way during the last part of the Emergent Mississippian period. On the summit stood a wooden

building, measuring 105 by 48 feet (32 by 15 meters) and perhaps 50 feet (15 meters) high.

From this perch atop Monks Mound, the elite rulers manipulated both the ceremonial and secular lives of the Cahokian people. Monks Mound and the other temple mounds were designed, by their sheer size and spatial arrangement, to emphasize the widening social distance between elite and commoners. Wearing clothing and jewelry befitting their elevated status, the new nobility—maybe 5 percent of the total population—literally towered over

In this reconstruction of the building of the sun circle at the American Woodhenge, the workers on the left excavate a post pit with a sloping ramp. The other workers insert a thirty-foot post into the hole.

everybody and everything. Townspeople supported their royalty, setting them apart from the population at large by social and political protocols. The elite and nonelite lived and died in spatial separation. The houses of the nobility were bigger than, and apart from, the dwellings of commoners.

Fronting Monks Mound is the Grand Plaza, a rectangular public space covering 46 acres (19 hectares). Recent remote sensing surveys have shown that this area was originally an undulating surface of ridges and swales. During the earliest phase of Cahokian development workers laboriously leveled and filled it in, removing massive amounts of fill from nearby pits. Excavations have turned up high-status and ritual debris discarded in these barrow pits, suggesting that such centrally sponsored public works projects involved the cooking and consumption of special ritual foods and medications.

Elite and Commoners at Mound 72

The nature of power and authority is dramatized by discoveries in Mound 72, a corporate mortuary facility for the Cahokian elite. It offers, says Thomas Emerson in *Cahokia and the Archaeology of Power* (1997), "the first blatant evidence for the existence of an elite with absolute control over ... multiple material and human resources (254)." According to Melvin Fowler, more than 260 people were buried here—nearly half of them as retainer sacrifices.

The Mound 72 burials were accompanied by a wealth of exotic manufactured goods, including huge quantities of marine shell artifacts, copper, and mica. One mass grave contains more than fifty young women; a nearby burial contains four beheaded and behanded males. After each major mortuary rite, the mound was enlarged by adding another mantle of soil.

At least one person in Mound 72—probably the paramount lord of Cahokia or a close relative—was laid to rest on a litter adorned with thousands of shell beads. Nearby were four hundred arrows, left in bundles or quivers; the arrowheads were manufactured of the finest exotic cherts. Mortuary goods also included nineteen polished discoidals ("chunky stones"), two bushels of uncut sheet mica, sheet copper, and the bodies of three men and three women, probably retainers who were sacrificed as part of the mortuary ceremony. These burials with associated human sacrifices and rich funerary goods signal the appearance of an elite whose superiority was buttressed by mythology and ideology.

Birger figurine, a kneeling woman on a round base.

Keller figurine, a woman kneeling on a base made from ears of corn or bundled reeds.

The Cahokian Stone Goddesses

As part of the FAI–270 Archaeological Mitigation Project, in conjunction with highway construction in the St. Louis area, Thomas Emerson supervised excavations at the small Mississippian site known as the BBB Motor site, roughly 2 miles (3.5 kilometers) northeast of Monks Mound. On this low ridge, isolated from the rest of the Cahokia complex, stood a small temple complex, mortuary area, and living quarters for priestly attendants. It dated from between 1050 and 1100.

Emerson's crew found two stone figurines buried in pits beneath this temple complex. Today known as the Birger and Keller figurines, they represent distinctive female deities that distinguish Cahokia from the rest of the Mississippian world. The stone goddesses provide a unique glimpse into the Cahokia belief system, a rare opportunity to see how ritual behavior and symbolic expressions were integrated into an overall framework of elite hierarchical power.

The Birger figurine represents a kneeling or squatting female on a circular base. One hand rests on the head of a feline-headed serpent, which bifurcates into gourd (or squash) vines that climb up her back and encircle her left shoulder. With the other hand, she is stroking (or perhaps tilling) the snake's back with a hoe. Emerson relates the serpent creature to the underwater monsters known from southeastern Indian cosmology, well documented ethnohistorically. These creatures were closely involved with water, rain, lightning, and apparently fertility. The agricultural motifs symbolize the natural process of regeneration.

Agricultural elements also appear on the Keller figurine, which depicts a woman kneeling on a base made of ears of corn, or perhaps bundled reeds woven into a mat. Her forehead slopes as though her skull had been deliberately deformed. Her long straight hair is pulled back, reaching her waist. From the basket that sits before her knees sprouts a plant stalk, probably maize, that grows through her right hand, sweeps along her back, finally attaching just above her ear. Emerson believes that the Keller figure symbolizes procreation of corn plants. It is an important finding because such agricultural motifs, although relatively common at Cahokia, are nearly absent elsewhere in the Mississippian world.

Both red stone figurines recall the Corn Mother tales told by many southeastern Indian groups. These figures, commonly associated with great monster serpents, initially provided sacred bundles of key plants such as maize, tobacco, and pumpkins to Native American people. They are also involved with transporting the dead between this and the other world. On the Birger figurine, Emerson points out, the basket and backpack motifs symbolize the containers used by the Earth Mother to transport not only crops but also bones and souls. Whereas the corn on the Keller figurine seems to directly represent Corn Mother imagery, the gourd vines growing from the Birger female's hands and body probably relate to a broader version of the Corn Mother tale.

In effect, the red goddesses were designed to deliver important concepts of the Cahokian cosmos to the rural masses. Through rites involving the procreation of key agricultural crops, fertility, and life forces, Cahokian society was drawn to a common symbolic core. Emerson argues that the ability to control fertility symbolism and ritual at rural centers gave the elite a way to maintain control over the hinterlands.

The Demise of Cahokia

Despite its impressive achievements, Cahokia was a relatively short-term affair. The years 1050–1200 were a time of consolidation, when the ruling elite at Cahokia controlled much of the American Bottom. This pattern persisted between 1200 and 1275, but the population of Cahokia dropped to only three or four thousand, and the period 1275–1350 was a time of minimal activity. Although mantles of earth may have been added to some of the mounds, there is little evidence of elite activity. This was probably a time of Mississippian population dispersal out of the American Bottom into secondary drainages and uplands. By 1300 Cahokia had fully disintegrated and would soon lie in ruins.

Nobody knows the exact cause of this decline. Over the centuries, the Cahokians must have depleted the available natural resources and degraded the agricultural potential of their floodplain. In addition, significant climatic change was occurring, with drier intervals cutting into the agricultural productivity upon which Cahokia depended. Some bioarchaeologists believe that dietary stress and increased disease levels might have played a role. Others emphasize the shifting horticultural balance: new, more productive strains of maize could have freed hinterland communities from reliance on the Cahokia-dominated redistributive network. These are all materialist explanations, stressing technology and environment as causal factors.

Some recent investigators, however, have shifted away from so-called "eco-functionalism" to explore the role of ideology in the rise and fall of Cahokia. For one thing, new research by Emerson and others indicates that the late Mississippians at Cahokia may have been as healthy as earlier populations. Additional radiocarbon evidence also suggests that Cahokia fell more rapidly than previously believed. Such an abrupt decline, coupled with continuing good health, seems to diminish the explanatory value of environmental decline.

In a 1992 article in the *Archaeological Papers of the American Anthroplogical Association,* Pauketat, rather than viewing ideology as "icing on a materialist cake," emphasizes the role ideology played in underwriting the political and religious authority of the chiefly elite at Cahokia. While not denying the relevance of materialistic factors, Pauketat, Emerson, and others underline the importance of belief systems that define an individual's or subgroup's relationship to the larger political entity.

In particular, they stress the importance of the emerging elite ideology in the urbanization and centralization process and the growing factional opposition that militated against centralized control and fostered instead a return to a more individualized universe. As Pauketat sees it, the emerging "stability is at the same time instability"—in the sense that control of this sort tends to be short-lived, due in part to clashes between elite rival factions (43). As the privileges of the paramount increase, so does the resentment from the have-nots. Limiting the paramount elite to high religious office may actually have diminished their nonreligious authority. The Cahokia elite benefited from their increased religious domination, with greater access to prestige items and increasingly grandiose mortuary rites. But the real-world consequences included a heightened degree of factionalism and rising social discord. Even at Cahokia's peak there were signs of social disintegration. Pauketat argues that, as the paramount elite became increasingly factionalized, they also became more isolated from practical political matters.

Once begun, Cahokia's final decline was absolute. Community-level organization had become increasingly decentralized, and rural districts enjoyed more local autonomy. With the emigration of certain elite groups out of the American Bottom and away from Cahokia, elite-controlled long-distance interactions withered.

Although traditional research has focused on a Cahokia "climax" or "peak of development," the new, idea-based perspective emphasizes instead the dynamic relationship between a series of small-scale chiefdoms during the Emergent Mississippian period. The resultant middle Mississippian "climax" is viewed as a short-lived period of "cultural homogeneity" in the American Bottom, marking a distinctive interval of cultural and probably political dominance by a single elite group.

Further Reading

A number of important works have recently been published on the archaeology of Cahokia. I particularly recommend *The Cahokia Chiefdom: The Archaeology of a Mississippian Society* edited by Alex Barker and Timothy R. Pauketat (Washington, DC: Smithsonian Institution Press, 1998); *Cahokia and the Archaeology of Power* by Thomas E. Emerson (Tuscaloosa: University of Alabama Press, 1997); *Cahokia and the Hinterlands: Middle Mississippian Cultures of the Midwest*, edited by Thomas E. Emerson and R. B. Lewis (Urbana: University of Illinois Press, 1991); *New Perspectives on Cahokia: Views from the Periphery*, edited by James Stoltman (Madison: Prehistory Press, 1991); *Mississippian Political Economy* by Jon Muller (New York: Plenum Press, 1997); *The Ascent of Chiefs: Cahokia and Mississippian Politics in Native North America* by Timothy R. Pauketat (Tuscaloosa: University of Alabama Press, 1994); and *Cahokia: Domination and Ideology in the Mississippian World* edited by Timothy R. Pauketat and Thomas E. Emerson (Lincoln: University of Nebraska Press, 1997).

Further Visiting

Cahokia Mounds State Historic Site (East St. Louis, IL; 6 miles [10 kilometers] east off of I–55/70 to SR 111 to Collinsville Rd.), a World Heritage Site, covers 2,200 acres (900 hectares) and preserves the sixty-five remaining Mississippian mounds.

Spiro

A.D. 850–1450

Mississippian culture

in Oklahoma

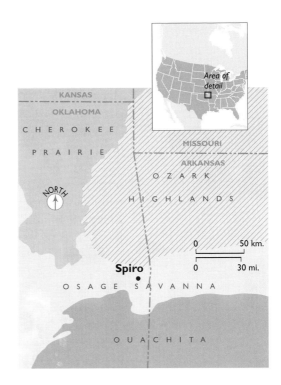

Spiro in eastern Oklahoma was one of the most impressive Mississippian religious and political centers in the southeastern United States. Named after the present-day community nearby, Spiro is situated on the first terrace and surrounding upland along the Arkansas River, near the Ozark highlands and the Ouachita Mountains, where the temperate forest begins to trail off into the vast expanse of the Great Plains.

The modern visitor can stroll around the 140 acres (60 hectares) of mounds and village ruins at Spiro, reminded of the magnificent Mississippian trade expeditions that once originated here. Unfortunately, little remains of the splendor of Spiro because relic miners carried it away. Vandalism and looting of important archaeological sites is a problem everywhere, but Spiro was hit particularly hard. What played out here has been called by archaeologist Jeffrey Brain, in a 1988 article in *Archaeology* magazine, "the single most disastrous destruction of a site in the history of American archaeology (20)."

Mining America's Past for Fun and (Especially) Profit

Lying in the swampy bottomlands and canebrakes of eastern Oklahoma, the Spiro mounds were among the last areas in the region to be cleared for farming. Although the property changed hands a number

This photo and newspaper gave an account of the vast fortunes made from looting Spiro.

of times, the spectacular mound complex remained unknown—at least to archaeologists—until late in the nineteenth century.

The Spiro mounds were first formally recorded in 1913 by local historian Joseph B. Thoburn. As rumors leaked out about the mounds and their suspected contents, a group of Great Depression-era "investors" banded together in 1933 to lease the rights to Craig Mound, the largest of the Spiro earthworks. Theirs was a strictly commercial enterprise, specifically directed at recovering and marketing the ancient artifacts buried in the mound. Calling themselves the Pocola Mining Company after a small town in Arkansas, they negotiated a two-year lease for a fee of fifty dollars per person.

The relic miners started by simply digging craters into the mound surface, looking for burials and isolated artifacts. But returns were relatively sparse and the mound was huge, so they soon hired some out-of-work coal miners, who brought with them a number of ingenious mining techniques. When the "investors" explained that the real treasure lay buried deep in the middle of Craig Mound, the coal miners showed them techniques of tunneling that allowed the looters to penetrate deep into the heart of the mound.

Here they struck paydirt. Craig Mound contained an unbelievable treasury of high-quality artifacts and relics, many of them decorated with distinctive Southern Cult iconography: stone effigy pipes, beads and pendants, ceramic vessels, copper plaques, axes, basketry, and cloth fragments. But particularly treasured were the thousands of spectacular marine shell cups and gorgets (throat ornaments) from Spiro. Some shell artifacts were deliberately broken, to create more marketable artifacts. Somebody set up shop immediately outside the tunnel entrance, selecting from among the artifacts as they were brought out. The new finds enriched and energized the diggers of Pocola Mining Company, who burrowed on relentlessly.

Spiro and the riches of Craig Mound soon burst onto the private art market. Literally thousands of Spiro artifacts were hawked to curio dealers, collectors of antiquities, and museums across America; some even found their way to Europe. Whole engraved shells sold for two dollars each, a considerable sum in Depression-era prices. When such items hit the antiquities market today, they fetch more than $10,000 each. There were preposterous tales about a mysterious "Chief's Room" with a secret entrance tunnel. A contemporary newspaper account called Spiro "the King Tut's tomb of Oklahoma," and tales of pearls by the bushel circulated wildly. One report claimed that Tiffany's in New York had valued a quart of Spiro pearls at half a million dollars. The demand for Spiro artifacts also led to a steady stream of fakes and false claims of new finds.

As professional archaeologists got wind of the bonanza at Spiro, they were initially skeptical. So many dazzling artifacts hit the market, in such great

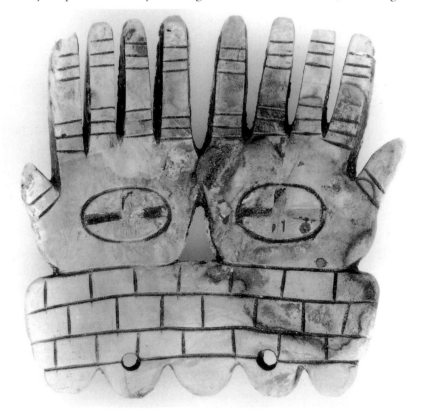

Paired shell hands from Spiro. This detailed marine-shell piece was either a pendant or headdress tied to a leather band. The hand, eye in the palm, circle, and cross suggest a funerary theme. The hands may represent those of a warrior or other revered member of Spiro society. The eyes could symbolize the sun god or the Spiro chief, his powerful earthly progeny.

Wooden feline, a six-inch (15-centimeter)-high puma excavated in 1884 at Key Marco, an island off Florida's western coast, showing distinctive Southern Cult markings

numbers, that many believed that the entire episode was a large-scale forgery. But as archaeologists eventually came to accept the authenticity of the finds, they were astonished by the unheard-of abundance and quality of the artifacts.

Public outrage over the vandalism at Spiro prompted Oklahoma legislators in 1935 to pass an antiquities law prohibiting such relic mining. One of the earliest state statutes designed to protect archaeological sites, the new Oklahoma law required individuals and institutions wishing to conduct excavations to apply for permission from the head of the Department of Anthropology at the University of Oklahoma, at that time Professor Forrest E. Clements.

The summer of 1935 was especially eventful. When the relic miners were notified that they would have to cease their now-illegal operations at Spiro, they scoffed at the new law. But under pressure from Clements, local law enforcement officers eventually forced the grave robbers to pack up and leave. Confident that he had solved the problem, Clements decided to escape the Oklahoma heat by accepting a summer teaching appointment in California. With Clements temporarily absent, the local sheriff was reluctant to enforce the new "city slicker" law. The Pocolo Mining Company's lease with the landowners did not expire until the fall, and the company swung back into operation throughout the summer, tunneling into the last remaining lobe of Craig Mound. There the diggers encountered—and destroyed—what they called the "Central Chamber," which contained hundreds of pounds of shell ornaments and beads, copper plates, feather cloaks, woven mats, and blankets that covered the elite buried there.

When Clements returned in the fall, he was shocked to see what had happened to the Spiro mounds. Digging furiously throughout the summer, the relic miners had tunneled the great mound through and through. The ground surface was littered with pieces of feather and fur textiles, broken pots, and fragments of intricate shell engravings. In a final contemptuous gesture, the gentlemen of the Pocolo Mining Company had set off dynamite charges inside Craig Mound, cracking the central core and causing still further damage to the remaining archaeological deposits.

After incessant delays over funding and requisite permissions, Clements finally was able to initiate responsible archaeological work at Spiro. The federal Works Project Administration joined with local Oklahoma historical and archaeological societies to protect Spiro from further depredations and to conduct excavations there. Detailed site maps prepared over five years plotted the locations of mounds, house foundations, trash deposits, and so forth. Although his finds could not rival those of the relic miners, who had essentially stripped the site, Clements did document the extent of earlier digging and refine the mound chronology.

Excavations by the University of Oklahoma and the Oklahoma Archeological Survey from the 1960s through the 1980s explored fifteen mounds (variously used for mortuary purposes, as house platforms, and to cover up previous dwelling areas) and a small settlement containing several rectangular houses. The excavators uncovered more than 750 burials dating between A.D. 850 and 1540. In 1971, the United States Corps of Engineers, which owns the Spiro Mounds site, recommended establishing an archaeological park. The park opened in 1978.

The Great Mortuary in Craig Mound

Given the unfortunate circumstances of Craig Mound's excavation, it is hardly surprising that considerable confusion arose. But thanks to recent research, particularly by James A. Brown, we have a better knowledge of how Craig Mound was constructed and what it contained.

Craig Mound was begun before 800, and elite burial continued there until about 1450. The mound actually consists of four conjoined units, or cones, constructed along a line 400 feet (120 meters) long and connected by low saddles. The northern or main cone originally stood 33 feet (10 meters) high at the highest point; the three lesser cones were about 15 feet (5 meters) high. The northern cone was built last, over a previously dismantled mortuary structure.

The large mound began as a series of small, conjoined burial mounds. These were accretional, built up in layers of mats, bark, burials, mortuary goods, and earth fill. A crematory 16 feet (5 meters) in diameter was also dug out; it had a 19-inch (48-centimeter) high rim and a flight of stairs leading inside. Apparently, several large mortuaries, also called charnel houses, were dedicated, used, then dismantled in this area.

The final ground-level charnel house, called by Brown "The Great Mortuary," was larger than the others, and the burial goods were left in place. The

Spiro site plan

One of the five hundred shell drinking cups from Spiro. This engraved conch shell has an intricate design, rendered in full form below. Shown is either a birdman deity or a human dressed in a falcon costume and mask, striking a ritual dancing pose.

floor, measuring 55 by 37 feet (17 by 11 meters), was covered with split cane oriented along the main axis of the mound. Innumerable caches of artifacts and disarticulated remains were placed here. The highest-ranking individuals were laid out in the thirteen cedar pole burial litters, heaped with valuable and unusual artifacts that included marine shell cups and gorgets, shell and pearl beads, fabric robes, copper plates, and basketry chests.

The cedar pole litters were themselves highly symbolic, intended to convey, on the shoulders of bearers, only the highest-ranking members of Spiro society. These litters were not only an agreeable means of transportation when everyone else had to walk but also demonstrated the superior status of those being borne. The smaller litters were hoisted by only two bearers. When brought into the Great Mortuary, they were packed with correspondingly modest allotments of grave goods. But the biggest litters were ten times larger, requiring four to eight bearers and containing huge quantities of funerary items. At least thirteen such litters were preserved in the Great Mortuary, doubtless brought there in formal and highly public processions.

When the Great Mortuary was closed, the contents were sealed beneath a large earthen platform mound, along with extra burial caches. Considerable air must have been trapped in and around the shell cups and other funerary items, accounting in part for their excellent state of preservation. The Great Mortuary contained the largest and richest single mortuary deposit yet uncovered in North America. This was the treasure trove that was mined out in the mid-1930s.

Life and Death at Spiro

The artifacts from Craig Mound make it clear that the dead buried there were persons of extraordinary import, probably a ruling elite that controlled the two hundred to three hundred ceremonial centers of the Arkansas River Valley. The overwhelming quantities of exotic prestige items—several thousand Gulf Coast shell cups, beads, and pendants; embossed copper plates, and additional ritual paraphernalia—make it clear that Spiro was inextricably linked to a pan-Southern iconographic network.

Spiro was one of the major Mississippian trade centers in Native North America. Its leaders controlled exchange networks that expanded onto the plains and across the southeastern United States. Mississippian people began living at Spiro as early as 900, but the fame of Spiro dates to the later phase, between 1200 and 1350.

Mississippian life at Spiro, as elsewhere, was played out in long-term political cycles. Local communities of thousands of people accepted the rule and leadership of a particularly effective chief. But upon his or her death, they fragmented once again into constituent communities. Competition for power and prestige—for control of key revenue-producing resources—became intense, both within local Mississippian communities and across regions. As political and social ranking proliferated, the Mississippian mind-set was increasingly reinforced by ceremony and sacrament. Beliefs expressed ancestral obligations; celebrated successful harvests, hunts, and warfare; and reinforced esteem for social leaders through elaborate mortuary ritual.

The Spiro grave goods tell us a great deal about ancient Mississippian lifeways. Ritual dress and ornamentation is vividly depicted on the engraved shell artifacts from Craig Mound and elsewhere at Spiro. Sometimes even specific individuals can be recognized. The falcon-impersonator, for instance, is known to have had a unique status in Spiro society. The falcon was a symbol of ferocious and courageous behavior, and the falcon-impersonator wears the forked-eye motif to symbolize this valor. It is one of the most distinctive and widespread motifs across Mississippian America.

Like many Mississippians, people at Spiro practiced a religion that recognized the importance of the seasonal cycle, conducting special ceremonies to ensure successful planting, bountiful harvests, prosperous life, and reverence in death. The mounds at Spiro supported large temples, mortuaries, and houses for the priests or chiefs. People living at Spiro witnessed hundreds of esoteric religious rituals, apparently centered around the death and burial of the elite members of their society.

Spiro society was graded according to social rank (if not outright social class), and mortuary patterning provides several clues about the social order. Most of the resident population was buried in local cemeteries and burial plots. Such funerary items as were interred with them consisted of utilitarian tools and implements used in everyday life: pottery vessels, arrowpoints, pipes, and items of personal adornment. By contrast, the higher class of Spiro society was singled out for special treatment. In some cases, the privileged were exhumed from the local cemeteries and reburied in the central mound complex. Higher-status individuals, found mainly in mortuaries associated with civic-ceremonial centers, were often subjected to postmortem manipulation. In fact, the rule seems to have been, the higher the status, the greater the manipulation. Those with relatively high status were partly disarticulated; only the elite were more completely disarticulated and cremated (in large specially constructed basins).

The lower-ranking regional elite, buried with great dignity and considerable grave goods, were placed in special areas of the Spiro site and in outlying centers throughout the Arkansas River drainage. The next level of the social hierarchy were buried extended, but disarticulated, inside specially constructed mortuary baskets and accompanied by significant quantities of grave goods. Those few individuals resting on litter biers in the Great Mortuary occupied the very peak of the Spiro social pyramid, as reflected by specialized funerary treatment and the lavish grave goods buried with them.

During its heyday, Spiro had become a vacant temple-town, where the elite and highest-ranking members of society were buried and venerated. Where did the rest of the people live? Surrounding villages, hamlets, and isolated

169

(top) Rattler pipe from Spiro
(middle) Well-preserved cedar mask
(bottom) Mississippian falcon-impersonator on shell gorget

The so-called Southern Cult was an iconographic network extending over much of the eastern United States during Mississippian times (about A.D. 1000–1600). Objecting to the somewhat shady meanings today associated with the term "cult," many archaeologists prefer the term Southeastern Ceremonial Complex. But whatever the name, it is clear that this massive pan-Mississippian network concentrated in three regional centers: Moundville in Alabama, Etowah in Georgia, and Spiro in Oklahoma.

The striking similarities in theme, motif, and medium imply more than simple trade networks; there was a higher degree of social interaction at work. The conch-shell gorgets and cups, the copper plates, the ceremonial axes and batons, the effigy pipes and flint knives found at Spiro and elsewhere contain a distinctive set of pan-Southern symbols. The forked eye, the cross, the sun circle, the hand and eye, the bilobed arrow, among others, suggest a shared symbol system that extended beyond the limits of any single Mississippian empire, spreading from Mississippi to Minnesota, from the Great Plains to the Atlantic Coast. In addition to small, "expensive" items, Southern Cult exchange may have involved critical subsistence resources such as food and salt.

Radiocarbon dating has shown that Southern Cult items have a longer temporal span than previously thought. They are today believed to be emblematic of several different Mississippian social institutions, including hereditary elites, military ranks, and priesthoods devoted to ancestral devotions. In this sense, the term Southern Cult generally refers to the entire body of Mississippian ceremonial art, implying at times a common Mississippian religion. Many of the representations of crosses, hands and eyes, sun symbols, serpents, woodpeckers, falcons, raccoons, and others, as well as ceramics modeled on animal and human forms, can be traced in the belief systems of postcontact Native Americans of the Southeast to their folk tales, myths, and religious observances.

Repoussé copper profile of a Mississippian man, recovered from excavations at the Spiro Mounds in eastern Oklahoma. Typical Southern Cult ornamentation includes the forked eye symbol (probably representing facial tattooing or painting), the distinctive hair bun, hair trimmed into roach style, and the copper ear plugs.

farmsteads have been found scattered throughout the Spiro hinterlands. The largest of these cover 10 to 20 acres (4 to 8 hectares), but most such settlements are smaller, covering less than 4 acres (2 hectares). They generally consist of a couple of structures and a small cemetery area.

One of these original Spiro houses has been reconstructed. The modern visitor can see how the walls were built of cedar posts set into the ground, with cane and grasses woven between. After clay was packed into and across the surface of the matting, a series of small fires ignited on both sides of the walls hardened them in the way a kiln fires ceramic pots. The roof was thatched with grass tied to rafters and cedar center posts.

Mississippian houses such as this probably sheltered an extended family of five to fifteen people. During the winter, they provided welcome warmth, and many domestic activities took place inside. During the summer these tasks were moved outdoors, and the houses provided cover from the frequent summer thunderstorms. Smoky smudge fires helped keep summer insects at bay. This house is typical of the Mississippian settlements, hamlets, and villages associated with various ceremonial centers that characterized most of the eastern Oklahoma valleys between 1000 and 1450.

Like so many Mississippian ceremonial centers, Spiro experienced a gradual decline in population and was abandoned about 1450. The reasons for its demise are not entirely clear. Many scholars believe that the modern Caddo and/or Wichita tribes are the modern descendents of the Mississippian people who once lived and worshipped at Spiro.

Further Reading

The archaeology of the Spiro site is presented in James A. Brown's *The Spiro Ceremonial Center: The Archaeology of Arkansas Valley Caddoan Culture in Eastern Oklahoma,* Volumes One and Two (Memoirs of the Museum of Anthropology, University of Michigan, 1996); *Pre-Columbian Shell Engravings from the Craig Mound at Spiro* by Philip Phillips and James A. Brown (Cambridge, MA: Peabody Museum Press, 1978); and *Contributions to Spiro Archeology: Mound Excavations and Regional Perspectives*, Oklahoma Archaeological Survey Studies in Oklahoma's Past edited by Daniel J. Rogers, Don G. Wyckoff, and Dennis A. Peterson no. 16 (1989).

For a general overview of early Mississippian archaeology, see *The Mississippian Emergence* edited by Bruce D. Smith (Washington, DC: Smithsonian Institution Press, 1990). Jeffrey Brain has also written an excellent introduction to the problem of looting at Spiro in "The Great Mound Robbery," *Archaeology* 41 (3) (May/June 1988): 19–28.

Further Visiting

Spiro Mounds Archaeological Park (Spiro, OK; 2.5 miles [4 kilometers] east of Spiro on SR 9, then 4 miles [7 kilometers] north) covers 140 acres (55 hectares) and contains a dozen Mississippian mounds. The interpretive center presents an introductory slide program, interpretive murals, and exhibits with striking artifacts including effigy pipes, ornaments, and trade items.

CHAPTER

FIFTEEN

———

Moundville

A.D. 1050–1500

Mississippian culture

in Alabama

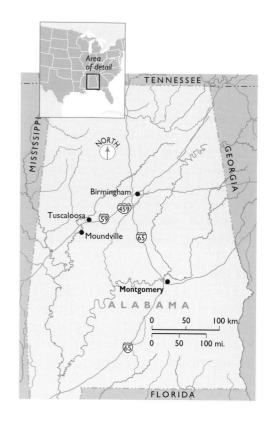

Moundville is one of the largest and most intensively studied Mississippian centers. Situated on a Pleistocene terrace overlooking the Black Warrior River in Alabama, Moundville consists of two dozen square and oval platform mounds, varying in height from 3 to 58 feet (1 to 18 meters). This compact, bounded settlement is fortified by a palisade that circumscribes the occupied areas on all but the northern margin, which is protected by the river bluff. Square-towered bastions were evenly spaced along straight sections of this palisade's curtain walls. Recent excavations reveal that the palisade trenches were renewed at least six times.

The largest mounds at Moundville were deliberately laid out along approximate cardinal directions, describing a single four-sided plaza. Some evidence suggests that this plaza area was artificially leveled before major mound construction. Larger mounds lacking burials alternate with smaller mounds containing human burials. In the center of the plaza stands Mound A, whose orientation is slightly different from that of the plaza-periphery mounds. North of the main plaza stand the flat-topped Mounds C and D, isolated on ridges formed by deep ravines; both contained numerous high-status burials. Toward the east and west, several other smaller mounds also lie outside the plaza area. Residential structures and rich habitation middens are scattered throughout the plaza-periphery mounds and inside the palisaded area.

History of Archaeological Exploration

Moundville was mentioned in Pickett's *History of Alabama* (1851), but the site was curiously overlooked in Cyrus Thomas's massive study, *Report on the Mound Explorations of the Bureau of Ethnology* (1894), despite the fact that the site had been mapped by agents of the Smithsonian Institution.

The site was first extensively explored by Clarence B. Moore, who piloted his flat-bottomed steamboat *Gopher* through the seemingly endless waterways of America's Southeast, excavating the major archaeological sites he encountered. In 1905, seven years before he worked at Poverty Point (Chapter 8), Moore paused on the Black Warrior River to examine the Moundville ruins. Anchoring opposite Mound C, Moore supervised several trained assistants and a crew numbering ten to fifteen. He prepared the first reliable site map, which remains useful today and is, remarkably, more accurate than a number of topographic surveys prepared decades later. Moore identified twenty-two mounds, and he set about exploring the summits of the large temple mounds, seeking cemeteries and hoping to unearth spectacular pieces of pre-Columbian art.

Moore concluded that Moundville was constructed and used before European contact and that it served as a major religious center. He commented on the clear-cut differences in social status between individuals buried in the mounds and those interred in off-mound cemeteries. He confirmed that the flat-topped mounds were built primarily to support buildings rather than as

Clarence Bloomfield Moore

Moore's stern-wheeler, from which he launched excavations at Moundville

Aerial photograph of
Moundville

burial mounds, although occasional, incidental interments occurred there. Moore further surmised from the varied art forms that the ancient people of Moundville worshiped the sun; in addition, motifs such as the plumed serpent and eagle suggested strong ties with contemporaneous Mexican civilizations.

A decade after Moore's departure from Moundville, the State of Alabama passed an antiquity act decreeing, among other things, that only citizens of Alabama were permitted to excavate or explore any archaeological site within the state. This states' rights approach to archaeology was clearly directed at Moore.

Although Moore's publications gave the impression that he had exhausted the archaeological potential of Moundville, quite the opposite was true. In 1927, Walter B. Jones, the director of the Alabama Museum of Natural History, initiated a new era of excavations at Moundville, at times drawing laborers from the Depression-era Civilian Conservation Corps federal relief program. Since the 1950s a number of smaller-scale excavations have been undertaken; some continue to this day.

Moundville occupies a pivotal position in the reconstruction of Native North American social complexity. In the 1970s, Christopher Peebles (then affiliated with the University of Michigan) pioneered the use of quantitative methods to explore the nature of social hierarchies at Moundville. Vernon Knight and Vincas Steponaitis have recently refined the regional chronology and established robust microchronologies of various mound and sheet midden deposits at Moundville. This revised chronology has significantly enhanced our understanding of Moundville's history. Steponaitis has refined the chronology for the site into five major phases that span the interval between 900 and 1650.

Moundville Before the Mississippians

The terminal Woodland occupation in the Black Warrior Valley appears mostly as large surface scatters identified during intensive archaeological surveys. In general, the latest Woodland period was characterized by endemic warfare and resource stress created by an expanding population. Some of the richer river valleys became unusually crowded, with large numbers of people crammed into closely spaced communities. But only a relatively modest late Woodland population seems to have lived in the Black Warrior Valley, where Moundville would eventually arise, and nothing in particular distinguishes these settlements from those of surrounding areas.

Recent excavations reveal that the subsistence base of these terminal Woodland settlements rapidly shifted from a reliance on wild foods to the production of maize at about 950–1000. There is also evidence of craft production at this time, especially the manufacture of shell beads. Throughout the Southeast, mortuary evidence strongly suggests that shell beads and pendants, commonly worn as jewelry and ornamentation on garments, had become the major standard of wealth, probably manipulated by rival community leaders who thus laid a foundation for competitive strategies that would soon characterize the elite classes at Moundville. But the Moundville site remained unoccupied until about 1050.

Mississippian Beginnings

The hallmarks of Mississippian culture appear during the initial occupation of Moundville: platform mounds, shell-tempered pottery, and distinctive wall trench architecture. Although people still grew and consumed native plant crops, corn made up about 40 percent of the average dietary intake. Most of the regional population lived in small farmsteads, although occupation on the Moundville terrace seems to have been unusually dense. Significantly, the only two mounds dated between 1050 and 1200 are found here; they probably served as foundations for residences of emerging elites. Although the mounds are fairly small, considerable labor was required to build them, suggesting that even at this early stage the elites living at Moundville were important in the regional political scene.

Knight and Steponaitis suggest that the leaders of this small-scale ranked society accumulated and distributed rare exotic materials as they expanded their spheres of political influence. These centralized authorities may also have been involved in interregional peace-keeping efforts.

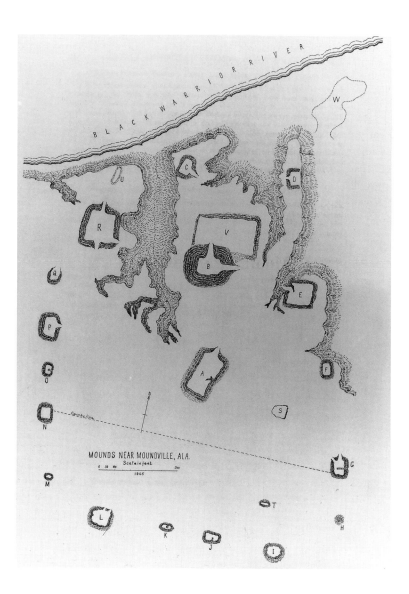

MOUNDS NEAR MOUNDVILLE, ALA.
Scale in feet
1905

Regional Consolidation

By 1250 Moundville had emerged as a paramount center, signaling the political consolidation of the surrounding region. New ceramic and radiocarbon evidence indicates that construction of both the protective palisade and all major mounds at this site began simultaneously around 1200–1250, defining for the first time the outlines of the central plaza. A clear-cut east-west symmetry was established, pairing residential mounds with mortuary temple mounds.

The character of Moundville changed at this point. People moved in great numbers from their unstructured farmsteads into the protected space inside Moundville, suggesting an overriding concern with security. The resident population numbered somewhat less than one thousand people, most of whom lived as nuclear families in square wattle-and-daub houses situated between the plaza and the palisade. Social space became quite formalized and ranked, with distinct public and residential areas carefully arranged around the quadrilateral plaza. Earlier houses were razed and the plaza margins leveled. One early mound (Mound X) was even "decommissioned" to make way for the eastern protective palisade.

At this point, maize agriculture provided about 65 percent of the Moundville diet. There is also evidence that the commoners were providing food tribute to support the Moundville nobility. A number of secondary satellite settlements arose, probably administrative centers that supervised local rituals and regulated the flow of tribute. Long-distance exchange flourished, as Moundville imported large quantities of nonlocal chert, greenstone, mica, copper, and marine shells. Specialized craftspeople worked these raw materials into flashy artifacts for social display.

Clearly Moundville had become the single overarching regional center of the Black Warrior Valley. Huge amounts of human labor leveled the plaza, built the mounds, and erected the protective palisade—reflecting the power amassed by the Moundville nobility. The site landscape suggests the presence of complex, ranked social statuses among the resident kin groups.

Sanctifying Moundville

Between 1300 and 1450—during what Knight and Steponaitis term the period of "paramountcy entrenched"—most of the Moundville population moved away to the farmsteads of the hinterlands. Several second-order mound sites sprang up, apparently to minister to the increasing rural population, which numbered perhaps ten thousand. Tribute, in the form of labor and agricultural produce, continued to flow inward to fill the needs of the nobility and their retainers who remained at Moundville. Clearly, the rulers at Moundville had distanced themselves symbolically and spatially from their followers.

Why did so many people leave?

Knight and Steponaitis think that Moundville may have been evacuated on direct orders from the elite, to sanctify the site and further separate rulers from commoners. Perhaps the decision was ecologically based. Maybe the large resident population at Moundville had severely depleted the soils and exhausted the wood resources. It is also clear that the protective palisade fell into disrepair and soon disappeared, suggesting that the threat of attack no longer existed. Another possibility is that the wall was simply unnecessary now that so few people lived inside.

The Moundville of 1300–1450 was a necropolis, a city of the dead. Many of the mounds in the primary center were abandoned, and mortuary activities shifted to cemeteries established in newly abandoned residential parts of

177
—

Various Southern Cult motifs on ceramics recovered by C. B. Moore at Moundville

Artifacts recovered by
C. B. Moore at Moundville:
(top) monolithic stone axe,
(center) stone effigy, and
(bottom) stone disk with
Southern Cult motifs

the site. With the resident population depleted and the paucity of burials in the outlying farmstead communities, it is clear that most of those buried at Moundville never lived there.

Collapse and Reorganization

People were buried at Moundville during the late fifteenth century, but on a diminished scale. Fewer individuals were buried in the off-mound cemeteries, and the elaborate tombs of the elite were no longer placed in the mortuary mounds. The focus had shifted from Moundville to outlying mound centers, which expanded during the sixteenth century. Nucleated villages sprang up throughout the Black Warrior Valley.

But this sense of freedom and self-sufficiency was short-lived, and the newly established secondary mound centers were in turn abandoned about the time Hernando de Soto and his party appeared in 1540. The lingering presence at Moundville disappeared by the end of the sixteenth century. The population of the Black Warrior Valley had shifted to a few moundless villages, without any trace of a hierarchy among settlements. Maize agriculture declined, replaced by a return to wild foods, and there are suggestions of disease and malnutrition. By 1650 the valley was becoming a depopulated buffer zone between warring proto-Creek and Choctaw peoples.

Some controversy exists regarding the demise of Moundville and its secondary centers. Christopher Peebles takes a long-term view, suggesting that a process of "social devolution" was responsible. Peebles believes that by the time of de Soto's dramatic entry, Moundville had weakened due to endogenous factors, realigning itself both internally and externally. Although he argues that population levels remained relatively constant, new alliances and trade networks developed. The quantity of imported exotics declined dramatically, thereby undercutting the ability of the elite to control the Moundville populace. Without a powerful ruling class, competition for resources intensified within Moundville society, ultimately leading to malnutrition and disease. According to this view, the demise of Moundville was not the product of external forces such as European intruders or their introduced diseases.

Charles Hudson, Marvin Smith, and Chester DePratter disagree, arguing that Moundville was brought down by direct military actions of the sixteenth-century European invaders and the Old World pathogens they inadvertently brought along with them. Hudson and his colleagues have painstakingly reconstructed the route followed by the de Soto expedition in late 1540. According to this reconstruction, the de Soto party arrived in the Moundville area on December 2. Whether or not Moundville is mentioned in the de Soto narratives remains a point of considerable debate. Based on documentary evidence left by expedition members, Hudson believes that de Soto encountered a unified, if somewhat diminished, chiefdom still operating in the

Death is an inescapable transformation—for the deceased and for those left behind. As the dead are separated from the living, they must be properly integrated into the world of the departed. Archaeologists believe that social ties between the living and the once-living reflect, in microcosm, the larger social relations of an entire society. Burial rituals thus provide a measure of an individual's worth at death. "Last rites" provide the archaeologist with fossilized clues about the terminal status of an individual.

Social ranking at Moundville formed a pyramid. At the peak was the chief of Moundville, a person of noble birth who was probably thought to be divine. On this paramount individual were lavished the most elevated emblems of status and rank. Slightly below the chief were those who enjoyed extraordinarily high status and considerable political authority. These individuals, all males, were buried in the large truncated mounds of Moundville, sometimes accompanied by infants and adult human skulls—perhaps retainers and kinsmen sacrificed for the occasion. Each mound contained only a limited number of these very high-status adults, whose grave goods included copper gorgets, stone discs, various paints, and assorted exotic minerals such as galena and mica.

Many of the elite grave goods displayed symbols of the so-called Southern Cult (see Chapter 14). Such very high-prestige items were buried with individuals of all ages and both sexes, and this is why many think that social status was assigned at birth. It looks as though one's social position in Moundville may have been inherited, then automatically passed along to relatives. This inference is reinforced by the fact that even infants and children too young to have accomplished anything very noteworthy in life were buried with lavish grave goods. Some archaeologists believe that these people were important strictly because of who they were, not what they did. Other archaeologists, such as Vincas Steponaitis and David Braun, would allow a greater role for achievement by extraordinary individuals at Moundville. It may be that lavish child burials, for instance, should not be automatically interpreted as signs of social status. In chiefdom-level societies, small children tend to reflect the social status of their parents, whether that status originated by birth, achievement, or both.

Some mounds also contained less well-accompanied (presumably lower-status) individuals, furnished for the afterlife with only a few ceramic vessels. These were probably second-order ritual or political officers who were buried in or near the truncated mounds.

At the base of the social order were adults and children buried in cemeteries near the mounds and in charnel houses near the main plaza. These villagers' graves also reflect their social position, largely a function of gender and age rather than inheritance. Here grave goods were assigned in a quite different manner. Graves contained pottery vessels, bone awls, flint projectile points, and stone pipes, which were distributed unevenly, mostly to older adults. These people achieved, rather than inherited, their social status, and prize artifacts went to the "self-made individuals." More than half of the Moundville graves were those of commoners, buried without any grave goods at all.

Moundville area. He also believes that although Moundville had indeed been weakened by internal political difficulties, it was the de Soto expedition that finished it off.

Knight and Steponaitis suggest that by the time DeSoto arrived, a relict Moundville elite still lived on the summits of the major earthworks. Although the paramount hereditary chief may have been recognized by various town chiefs, no real political power was involved, as among the present monarchy of England.

Further Reading

Vernon James Knight, Jr., and Vincas P. Steponaitis have recently published a major resynthesis: *Archaeology of the Moundville Chiefdom* (Washington, DC: Smithsonian Institution Press, 1998). Another important recent source is the republication of Moore's classic monograph in *The Moundville Expeditions of Clarence Bloomfield Moore,* edited with an introduction by Vernon James Knight, Jr. (Tuscaloosa: University of Alabama Press, 1996). See also *Ancient Chiefdoms of the Tombigbee* by John H. Blitz (Tuscaloosa: Univesity of Alabama Press, 1993), and *Towns and Temples along the Mississippi* edited by David H. Dye and Cheryl Anne Cox (Tuscaloosa: University of Alabama Press, 1990).

Further Visiting

Moundville Archaeological Park (Moundville, AL;.5 miles [1 kilometer] west of SR 69 at the edge of town), one of the country's best-preserved prehistoric mound groups, lies on the south bank of the Black Warrior River. Graded paths climb to the mound summits, once topped by ceremonial houses and chiefs' lodges; a reconstructed temple dominates one mound. The Moundville Archaeological Museum is also housed here, providing Alabama's teachers with up-to-date material about Native Americans and offering educational opportunities about Native Americans for both students and the general public.

SIXTEEN

Iroquoian
Archaeology

A.D. 1300–1650

Iroquoian tradition

around London, Ontario

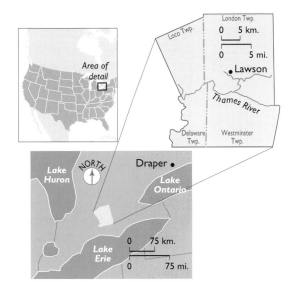

As seen in the preceding chapters, Mississippian settlements clustered along the rich, low-lying flood-plains of the Mississippi River and its major tributaries. The Mississippians, perhaps the most socially complex societies north of Mexico, erected the largest archaeological sites in North America, organizing labor pools for public construction. Their societies were socially stratified into elites and commoners, their enterprise fueled by high-yield maize agriculture.

But in areas where intensive Mississippian-style agriculture could not be practiced many Woodland societies continued their mixed economy of hunting, fishing, and harvesting both wild and domesticated crops. The Woodland farmers of the Northeast planted a single corn crop each year, interspersed with beans, squash, and sunflowers. A village of three hundred people would need about 100 acres (40 hectares) in the first year of production; this total would double after a decade or so, once productivity declined and new fields were opened up. Once the fields were exhausted, perhaps in a couple of decades, the village moved to a new location, often only a few miles away. This chapter and the next explore what took place in Woodland societies beyond the borders of the Mississippian phenomenon.

This chapter is concerned with the ancestors of the Iroquoian-speaking people who populated the valleys, mountains, and flatlands of New York and southern Ontario. Linguistic evidence suggests that the ancestors of the historic-period Iroquoians probably entered northeastern North America sometime between A.D. 700 and 1000, breaking off from their southern relatives and disrupting neighboring Algonquian-speaking residents of the area, whose territory spread from Nova Scotia to North Carolina. In this usage, the term "Iroquois" refers to those tribes, mostly in New York, that eventually joined the seventeenth-century League of the Iroquois. "Iroquoian" refers to both these and other Iroquois-speaking groups living in northeastern North America.

The Ontario Iroquoian Tradition

The Ontario Iroquoian tradition documents the origins of three historic-period Iroquoian tribes: the Huron, Petun, and Neutral. On the basis of archaeological and ethnohistoric evidence, this tradition is conventionally divided into early (A.D. 1000–1300), middle (1300–1400), and late (1400–1650) periods.

By about 1000, the Iroquoian ancestors had spread out to Lake Huron and Lake Erie in the west and to the Hudson River Valley and Canada's Gulf of St. Lawrence in the east. These farmers took up a characteristic upland horticultural system based on improved strains of maize, beans, and squash. Tobacco was also present, but not in great quantity until after about 1350. Most archaeologists believe that cultivation was conducted primarily by women. Early phase occupations were year-round settlements of longhouses protected by wooden palisades. Small, seasonal hamlets were established not far from these

Location of late Ontario Iroquoian people in southern Ontario

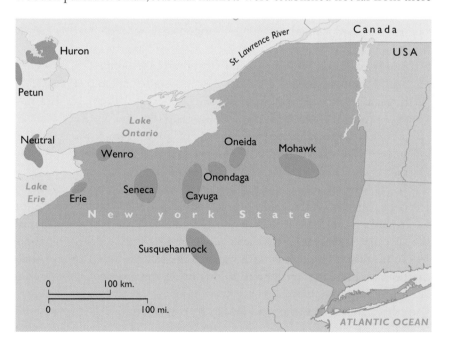

villages to facilitate exploitation of specific resources such as chert quarries, fishing grounds, and so forth.

Longhouse reconstructions at Crawford Lake

This initial Iroquoian occupation has been intensively studied by archaeologists affiliated with the London Museum of Archaeology. Particularly important have been excavations in the Crawford Lake area on the Niagara Escarpment in Milton, Ontario. The Crawford Lake Study Area covers an approximately 20-mile (32-kilometer) radius around Crawford Lake. Within the study area are numerous early Ontario Iroquoian villages, hamlets, and special-purpose sites dating from 1000 to 1300, although none is close to the lake. The Crawford Lake area is an excellent place to study the Iroquoian occupation because it is located along the historic-period Huron-Neutral boundary.

The archaeological significance of this area first surfaced as a result of paleoenvironmental research conducted by geographer Roger Byrne of the University of California at Berkeley. Crawford Lake is meromictic, meaning that, because of its small surface area and extreme depth, water in the upper half of the lake does not mix with water in the lake bottom. This condition creates a very low oxygen content below the 40-foot (12-meter) level. Bacteria are almost completely absent, so that organic objects such as pollen, seeds, and leaves are preserved in the undisturbed sediments at the bottom for centuries.

Byrne discovered that the sediments of Crawford Lake are stratified into a sequence of what are called varves, in which light sediments from the summer runoff alternate with dark sediments deposited during the winter. From a core sample of these sediments, he could precisely date the annual strata by counting the varves, just as a dendrochronologist counts the rings on a tree stump. The Crawford Lake sediments preserved a remarkably fine-grained paleoenvironmental record spanning the last four thousand years.

Byrne detected two major episodes of disturbance. One in 1846–1851 was attributed to Euro-American settlement of the area. The other indicates deforestation from 1290 to 1610. Particularly in the varve sequence spanning 1434–1458, Byrne isolated evidence of forest clearing: grass, corn pollen, seeds of portulaca (a weedy plant found in corn fields), charcoal, and increased sedimentation. The lake core thus documents a settlement from 1434 to 1459, probably the terminal occupation at Crawford Lake. Earlier occupations probably date to a century earlier.

Knowing that windblown corn pollen can travel only a short distance from the parent plant, Byrne predicted that archaeological sites and their associated maize fields could not be far away. William Finlayson and his colleagues eventually located and tested more than two dozen Iroquoian sites in the Crawford Lake area. These excavations have established the basic ceramic chronology for this frontier area.

The first Iroquoian-speaking group arrived in the Crawford Lake area about 1300. The Crawford Lake site was not a full-time village, but rather a seasonal hamlet. The initial occupation of Iroquoian speakers contained four or five longhouses. It was abandoned then reoccupied by a later group, which built three or four additional longhouses, some superimposed over the original houses. It seems likely that the people of one of these settlements probably cleared and planted the fields that produced the corn pollen deposited in Crawford Lake.

Although excavations at Crawford Lake village have turned up human war trophies, suggesting that the Crawford Lake group was engaged in warfare, Finlayson did not locate a full palisade protecting the village. The Crawford Lake settlement may have been removed from frontline warfare, but it also seems likely that, in the event of local skirmishes, residents may have retreated to the safety of their neighbors' larger, palisaded villages.

Longhouse Archaeology: What Does It Mean?

Nine conspicuous longhouses have been located at the Crawford Lake site. They range in length from 80 to 150 feet (25 to 45 meters). Today visitors to Crawford Lake can see one completely reconstructed longhouse and another full-size skeleton framework. The reconstructed Crawford Lake village apparently lies in Huron territory, although a number of contemporary settlements appear to be Neutral sites.

The distinctive longhouse became the hallmark of Iroquoian archaeology throughout southern Ontario and New York State. As time passed, these longhouses would grow to immense proportions, some of them longer than a modern football field. Some longhouses elsewhere in southern Ontario reached a length of 400 feet (120 meters) containing up to two dozen hearths, although these are not known from the Crawford Lake area.

The typical longhouse was barnlike, with windowless walls framed by upright elmwood saplings spaced at about 3-foot (1-meter) intervals, faced and roofed with elm bark. The standard unit of measure appears to be the armspan. Mohawk longhouses had four parallel rows of posts (two outer walls and two along the aisle) set roughly an armspan apart, creating houses about 20 feet (6 meters) wide. Dean Snow thinks that the variability in house width is probably due to the different armspans of those in charge. Primary posts are set at 7-foot (2-meter) intervals, with secondary posts added in between for strength, creating the observed 3-foot (1-meter) intervals.

Carefully placed rooftop vents allowed smoke to escape from the dim interior. Outdoor fires were used in times of good weather. During periods of heavy rain or snow, the smokeholes could be partially closed by shutters. Along each sidewall were berths on raised platforms, separated for privacy by side curtains. Whole families slept here, snug in their fur bedding. Above their heads were deep shelves for storage of baskets and pots, weapons, mats, spare skins, and antlers. From each overhead pole and rafter hung drying foods: corn, squash, apples, and herbs.

185

Draper site plan view

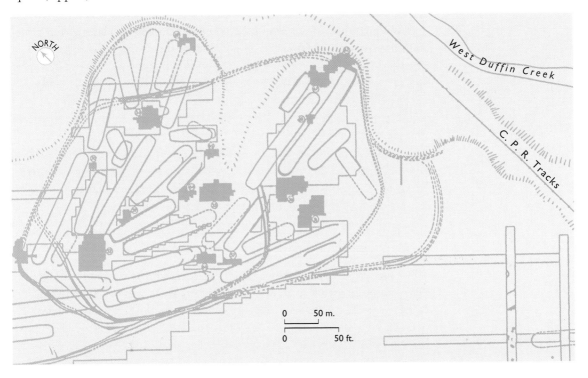

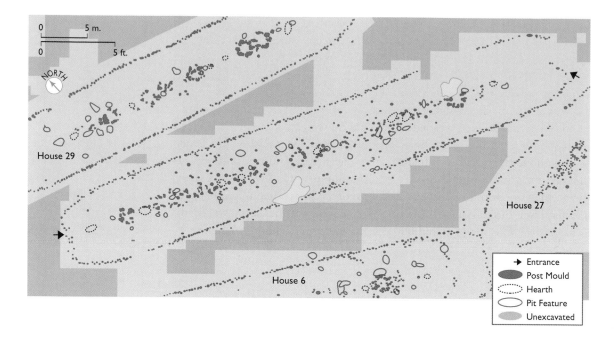

Draper site, House 38

0 5 m.
0 5 ft.

NORTH

House 29

House 27

House 6

Entrance
Post Mould
Hearth
Pit Feature
Unexcavated

The longhouses were built side by side to create distinctive Iroquoian villages. The Draper site northeast of Toronto, Ontario, had at least thirty-seven such longhouses dating to between 1450 and 1500. The Draper longhouses, housing two thousand people, covered an area of 20 acres (8 hectares) and were protected by wooden palisades. Small wonder that European explorers, encountering such villages during the mid-seventeenth century, would apply the term "castle" to the Iroquoian settlements.

Most archaeologists draw upon ethnohistoric Iroquoian settlement patterns and social structure to read meaning into longhouse archaeology. The Iroquoians determined their kin relations through the female line. In each longhouse, under a single roof, lived related women and their children, members of a single matrilineal clan, presided over by the clan mother, the oldest and most esteemed woman in a longhouse. This means that families living together within a longhouse were typically members of the same clan, all tracing their descent to a common female ancestor through the mother's line—and, of course, the in-marrying husbands who belonged to another clan. With the passage of generations such distinctions became blurred, so that all those living inside a single longhouse were not necessarily related by blood. Over the door of the Iroquoian longhouses were depictions of the clan's first ancestors, after which the clans were named: the Eagle Clan, the Heron Clan, the Wolf Clan, the Beaver Clan, and so forth.

Children of both sexes traced clan affiliation through their mothers' sides. A young man grew up in the longhouse of his mother and left home upon marriage to live in the longhouse of his wife. Except for their weapons, clothing, and personal possessions, everything in Iroquoian society, including the longhouse itself, belonged to women. Even though a husband might be shabby, his wife was expected to be well-dressed and respectable. Should a male kinsman be killed in war, a woman was entitled to demand an enemy captive in compensation. She was free to torture or kill him as she pleased. The clan mothers appointed and dismissed the councillor-chiefs.

Women controlled Iroquoian village society for solid, practical reasons. As hunters and traders, men traveled extensively. An ambitious military campaign meant that war parties might be gone for days or months. Although a handful of younger men were left behind to defend the longhouses, most aspects of community life, from nursing to childcare, from planting to harvesting, were the women's responsibility. Over the course of a year, women might bring in a half-million bushels of corn and tons of beans, squash, and sunflower seeds. What they did not prepare immediately, they stored for the future in underground granaries. These ancient matrilineal clans became the building blocks of later Iroquoian political institutions.

Clan affiliation extended far beyond individual longhouses, and clan membership could usually be traced for great distances, across tribal boundaries. Such clan relations eased long-distance travel and communication; visitors were always welcomed into the longhouse of their mother's clan. But because all clan members were considered to be blood relatives, marriage to even distantly related clan members was rigidly prohibited.

Fanciful Huron deer drive

Perhaps, as among some historic-period groups, there were distinct chiefs for peace and war. Those in charge of peacetime watched over domestic disagreements, marshaled public works projects, and monitored the ceremonial life of the village. For the Ontario Iroquoians, each community celebrated a formalized Feast of the Dead—when a new village was established, the remains of the deceased were disinterred and taken along with the community for reburial in the new locality. Obligations of the war chiefs, whose responsibilities were more restricted, extended over raids and prisoners. Little real power was involved in either office.

Archaeologists are inclined to extend aspects of this reconstruction back perhaps one thousand years in time. But as Bruce Trigger has emphasized, such inferences are possible only through a heavy reliance on the early ethnohistorical data, which is often presumed to represent Iroquoian culture before European contact. Today it is clear that even the earliest European observations, dating from the early 1600s, describe societies already involved in change as a result of direct or indirect contact with the European world. Archaeological investigations of early contact-period sites demonstrate that European trade goods had begun to reach southern Ontario nearly a century before actual contact.

So our understanding of Iroquoian residence and descent patterning, even during the historic period, contains many uncertainties, and the extension of this pattern into the archaeological past will never be entirely reliable. Archaeologists must particularly guard against stereotyping the archaeological remains in light of ethnohistoric evidence.

Lawson: The Late Phase of the Ontario Iroquoian Tradition

The Lawson site is located on a flat plateau overlooking the junction of Snake Creek and the Medway River in the city of London, Ontario. Occupied during the latter half of the fifteenth century, Lawson typifies the terminal phase of the Ontario Iroquoian tradition.

Tens of thousands of artifacts have been recovered at Lawson, both in scientific excavations and by looters who have collected artifacts here for a century. The Lawson site was first excavated in the early 1920s by William J. Wintemberg of the Victoria Museum, which became the National Museum of Canada and is now known as the Canadian Museum of Civilization. The Museum of Indian Archaeology—now the London Museum of Archaeology, affiliated with the University of Western Ontario—has excavated at the Lawson site annually since 1976 under the direction of Robert Pearce. In 1983 the London Museum of Archaeology was built next to the reconstruction of the Lawson site.

To date, all or parts of fifteen longhouses have been excavated at Lawson. They range from 40 to 110 feet (12 to 33 meters) in length. Most contained central hearths, several refuse-filled storage pits, storage cubicles, and internal

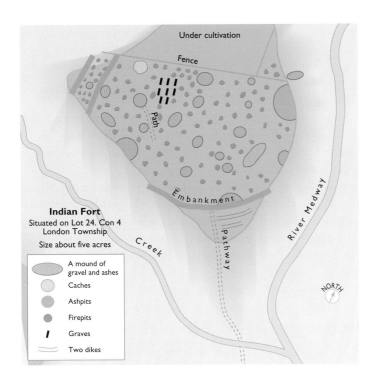

Indian Fort
Situated on Lot 24. Con 4
London Township

Size about five acres

A mound of gravel and ashes

Caches

Ashpits

Firepits

Graves

Two dikes

partition walls that divided smaller storage rooms from the main living areas. A single entrance was located at the west end of each house. Benches ran intermittently along both sides of the house. The longest house, which is also wider than the others, has broad sleeping benches; it contains fewer refuse-filled pits, but a number of small pits contain distinctive artifacts such as deliberately buried still-articulated beaver feet and an elk antler war club. Pearce has interpreted this large longhouse as the "chief's" house; a smaller, adjacent house may also have been used by this same Iroquoian leader.

By the time Lawson was established, warfare had emerged as a major factor in Iroquoian life, and several lines of archaeological evidence demonstrate the degree to which the Lawson villagers were preoccupied with defense. The village location was selected for its defensibility, situated on a plateau naturally protected by ravines on three sides. Such a strategic position is characteristic of Iroquoian villages of this time, which were built away from major waterways, a mile or two inland, on high ground naturally protected against enemy attack.

The original 4-acre (1.5 hectare) settlement at Lawson was protected by six rows of wooden palisades, built along the part of the site that was not naturally protected by ravines. In historic-period Iroquoian villages, such palisade timbers sometimes reached a height of 20 feet (6 meters). In addition there were defensive earthworks, ditches, and a complex entrance maze. Near one of the several lookout platforms, archaeologists found clusters of fist-size limestone cobbles stacked inside the stockade, doubtless stockpiled to be hurled down against enemy attackers. During times of war, villagers would place enemy scalps on large poles above the palisade entrances to frighten attackers.

Other signs of warfare are abundant at Lawson. An early excavator at Lawson reported finding more than five hundred human bones, of both adults and children, scattered across the site. Several pieces of human bone show signs of having been cut, cooked, and split open for marrow extraction. In many cases, human bones are found in refuse deposits generally associated with bones of food animals. If these are signs of cannibalism, the practice may correlate with increasing hostilities created by the sudden expansion of the Ontario Iroquoians. Although cannibalism would survive into the historic period, the practice appears to have peaked in the mid-sixteenth century.

Several human bones were modified into tools and artifacts, including five human skull gorgets, perhaps war trophies. Among the historic-period Iroquoians, warfare was considered a way to avenge the killing of kinfolk. Sometimes captives, particularly women and children, were brought back to replace deceased victims. More frequently male prisoners of war were killed in a flamboyant ceremony as a sacrifice to the sun. Cannibalism also became an integral part of a cult in which male prisoners of war were tortured to death. When the enemy was killed outright, the head or scalp was brought home as a trophy of war.

Robert Pearce, who excavated at Lawson in the late 1970s, suggests that subtle artifact patterning also documents a pattern of warfare. A distinctive kind of pottery called Parker Festooned, found in some quantity here, is

These longhouse reconstructions at the Lawson site are adjacent to the London Museum of Archaeology.

known to have been manufactured in villages to the west of Lawson. More than 80 percent of the stone tools are made from the Kettle Point chert, which occurs only in limited outcrops in this same westward area. Pearce hypothesizes that Lawson site occupants made frequent trips to Lake Huron to obtain Kettle Point chert. On these expeditions, they may have fought with the native people to the southwest, who in turn carried out reciprocal attacks on the Lawson site, necessitating the numerous defensive measures documented there. It may also be that the Parker Festooned pottery vessels made their way to Lawson either through exchange or the capture of enemy women and their possessions. All this indicates a pattern of long-standing warfare with Native peoples to the west.

Archaeological evidence throughout southern Ontario suggests that the previously dispersed independent Iroquoian villages began to group together for mutual protection during a time of increased warfare. Building long-houses closer together might have postponed the need to build protective palisades, but not for long. Lower-level alliances seem to have proliferated until Iroquoian populations clustered in a few select areas separated by large no-man's zones.

Warfare would increase throughout southern Ontario in response to conflicts over the fur trade, pitting the Huron and French, and indirectly the Petun and Neutral, against the New York Iroquois and Dutch. The Neutral were eventually defeated in the 1640s by the Iroquois from New York State, who also defeated and dispersed the Huron and Petun a few years later. Some of these people survive today, but the Neutral were totally eradicated. By 1651, the Native cultures of southern Ontario had been absorbed into other tribes and no longer existed as tribal entities. Southwestern Ontario became a no-man's land until Ojibwa groups moved into the area in the early eighteenth century. After the American Revolution, New York State Iroquois groups entered the area. Muncey and Oneida people settled along the Thames River during the late eighteenth and early nineteenth centuries, and they continue to live there today.

Why Iroquoian Warfare?

Several theories have been offered to explain the prevalence of warfare in Iroquoian culture. Some researchers suggest that the maize-dependent populations were forced into deadly competition by drought conditions during the fourteenth century—the same Great Drought that struck southwestern Pueblo societies so severely (see Chapter 11). Competition for prime farm land may have intensified to the point that Iroquoian people were forced to leave home in search of more productive territories.

Others, questioning the evidence for such major ecological changes, have suggested that sociocultural factors may better account for Iroquoian warfare. John Witthoft and Bruce Trigger have tied the rise of warfare to an increased emphasis on personal status and prestige. They assume that in middle Woodland times, male hunting skill was a primary test of personal skill and stature. But the analogy to early contact Iroquoian society, discussed above, suggests that the transition to a horticultural economy brought with it an elevated status for women, causing Iroquoian men to feel threatened by the growing importance of female-controlled horticulture. Witthoft and Trigger have suggested that males may have turned to increased raiding and warfare to reassert their male dominance in an era of increased pastoral sedentism.

Others have trouble with the jealousy argument. Matrilineal systems break up potentially competitive fraternal groups, uniting them and refocusing their hostilities onto outsiders. The Iroquois, Hurons, and Neutrals took this a step further by using the clan system as a foundation for confederations that eliminated aggression between confederated nations and intensified it between confederations and unconfederated nations in the region.

Certainly among the contact-period Iroquoians, warfare was not directed toward acquiring land, raiding for food, or controlling the fur trade. Iroquoians of the seventeenth century sought blood revenge—a way for young warriors to avenge previous killings and enhance personal prestige. It is not hard to imagine that a pattern of small-scale raiding and capture of prisoners could have escalated into a state of endemic hostility and outright warfare.

Further Reading

The most important publication on the archaeology of the Lawson site is the new four-volume series *Iroquoian Peoples of the Land of Rocks and Water, A.D. 1000–1650: A Study in Settlement Archaeology, Volumes I-IV* by W. D. Findlayson (Ontario: London Museum of Archaeology, 1998). See also Findlayson's "The 1975 and 1978 Rescue Excavations at the Draper Site: Introduction and Settlement Patterns," *National Museum of Man Mercury Series, Paper 130* (Ottawa: Archaeological Survey of Canada, 1985), and W.J. Wintemberg's *Lawson Prehistoric Village Site, Middlesex County, Ontario* (Ottawa: National Museums of Canada, 1939).

For general considerations of Iroquois archaeology, I especially recommend Bruce G. Trigger's *Natives and Newcomers* (Kingston, Ontario: McGill-Queen's University Press, 1985) and *The Children of Aataentsic: A History of the Huron People to 1660* (Kingston, Ontario: McGill-Queen's University Press, 1987); see also The Ontario Iroquois Tradition by James V. Smith (Ottawa: National Museums of Canada, 1966), and *Foundations of Northeast Archaeology*, edited by Dean R. Snow (New York: Academic Press, 1981).

Further Visiting

Lawson Prehistoric Indian Village is located adjacent to the London Museum of Archaeology in London, Ontario. Crawford Lake Indian Village, including a reconstruction of this fifteenth-century Iroquoian site, is located in Milton, Ontario.

Knife River

A.D. 1300–1800

Plains Village tradition

in North Dakota

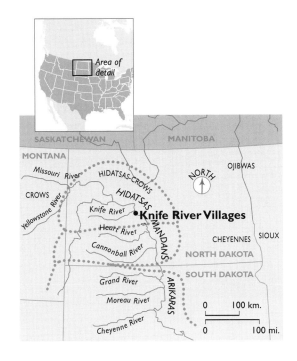

The Knife River Indian Villages National Historic Site, just north of Bismarck, North Dakota, contains an unbroken record of five hundred years of settled village life from the time when people arrived in the region around A.D. 1300. The Knife River "site" actually consists of three visitable earthlodge villages. Big Hidatsa (1600–1845) was the largest of the fortified villages along the Knife River, with 113 lodges. Sakakawea village (1790s–1834) was located directly on the west bank of the Knife River, with three or four dozen lodges; it was burned to the ground by Sioux raiders. The Lower Hidatsa village (1525–1780) contains fifty-one recognizable earthlodges. These villages, clustered along the banks of the Knife and Missouri rivers, were home to the Hidatsa people, who numbered perhaps three thousand to five thousand people.

The Knife River villages offer a unique experience. First-time visitors sometimes miss the subtleties of the rolling prairie landscape in front of them. But once the eyes adjust, a world of buried history appears: hundreds of huge earthlodge and cache pit depressions, discarded artifacts scattered on the surface, trash middens wedged between the houses, and linear fortification ditches, dug long ago to protect the villagers from attack.

194

Most subtle of all are the travois trails, distinctive V-shaped tracks worn into the prairie sod over the centuries. Before Euro-American contact, Plains Indians commonly hitched their dogs to travois, devices consisting of two long poles bound together by rawhide rope to form a platform. The back ends of the travois poles dragged on the ground behind. Each dog pulled an individual load. When horses arrived, they were fitted with larger travois, making it possible to transport full-sized tipi poles and other cumbersome household goods.

Dozens of travois trails are evident along the margins of the Knife River villages. These parallel drag marks are four to eight feet apart and deepened from repeated use. They silently show where Hidatsa people followed the tracks of their ancestors, moving between summer and winter villages and sending hunting parties further up the Missouri River. The first Euro-American explorers learned to follow these trails, the best route for traveling from village to village.

"A Very Curious and Pleasing Appearance . . ."

The earliest written account of these people comes from the French explorer Pierre Gaultier de Varennes, who headed a trading and exploring expedition along the upper Missouri River in the late 1730s. Then came eighteenth-century Euro-American fur traders. As experienced traders themselves, the Knife River villagers welcomed the newcomers into their earthlodges. Throughout the nineteenth century, they hosted a steady stream of colorful visitors, including members of the Lewis and Clark expedition (1804–1806), George Catlin (1832), and Prince Maximilian and Karl Bodmer (1833–1834). These visitors produced journals, maps, paintings, and drawings that recorded their detailed impressions of the tightly clustered, palisaded communities.

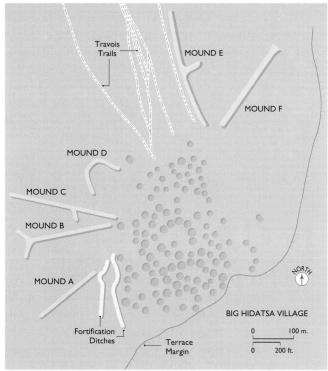

Paired views of Big Hidatsa
Village: (above) aerial shot
and (left) map of same

Travois
Trails

MOUND E

MOUND F

MOUND D

MOUND C

MOUND B

MOUND A

NORTH

BIG HIDATSA VILLAGE

Fortification
Ditches

Terrace
Margin

0   100 m.

0   200 ft.

All visitors, past and present, have been drawn to the sturdy earthlodges, admirably suited to the challenging environment and built to house from ten to thirty people. According to Catlin's account of his 1832 visit, as quoted in *People of the Willows* (1991),

> There are . . . dwellings [all] about me and they are purely unique— they are all covered with dirt. . . . The groups of lodges around me present a very curious and pleasing appearance, resembling in shape . . . so many potash-kettles inverted. On the tops of these are to be seen groups standing and reclining, whose wild and picturesque appearance it would be difficult to describe. Stern warriors, like statues, standing in dignified groups, wrapped in their painted robes, with their heads decked and plumed in quills of the war-eagle; extending their long arms to the east or the west, the scenes of their battles, which they are recounting to each other.
>
> On the roofs of the lodges, besides the groups of living, are buffalo skulls, skin canoes, pots and pottery; sleds and sledges—and suspended on poles, erected some twenty feet above the doors of their wigwams, are displayed on a pleasant day, the scalps of warriors, preserved as trophies; and thus proudly exposed as evidence of their warlike deeds. (14)

The lodges belonged to the Mandan and Hidatsa women, or, more properly, to their clans. Women were the primary architects, builders, and maintainers of their homes. Although they looked like huge earthen mounds, the lodges were mostly wooden, and each required the cutting and transport of perhaps 150 trees. A woman would first pinpoint the central point of her house and then measure a circle on the ground to define the lodge's perimeter. Most houses were between 35 and 40 feet (10 and 12 meters) across, although some Mandan ceremonial lodges measured up to 70 feet (21 meters) in diameter.

Enlisting the help of her men, the woman then erected a sturdy central framework of four uprights, each standing 10 to 15 feet (3 to 5 meters) high, to support the weight of the house. The uprights were connected by thick horizontal stringers to form the outer wall of the house. Next came the rafters, radiating outward like spokes in a wheel from a central opening that served as a smoke hole and window for daylight. No fasteners or nails were used, the entire superstructure being held together with a network of natural tree forks, notches, and wedges. Split logs were laid against the outer stringers to complete the wooden dome. Overlying layers of willows, bundles of grass, and tamped earth created insulating walls 2 to 4 feet (60 to 120 centimeters) thick. Depending on local tradition, the roof could be flat-topped or rounded. The rooftop was important because villagers loved to sit atop their houses, lounging with dogs, telling and retelling stores. Some kept their skin-covered bullboats up there, stored upside-down over the smoke holes to keep out the rain and prevent puppies and youngsters from falling through.

Today, visitors to Knife River can view the remains of nearly two hundred earthlodges. Only a handful have been tested archaeologically, but all have been explored using modern remote-sensing technology. Researchers have used low-altitude aerial photography to map each house depression and its relationship to the trails, linear ridges, and other features that crisscross the sites. False color infrared aerial photography was particularly useful for documenting subtle vegetational changes, which in turn reflect underlying archaeological and geological features.

Geomagnetic surveys provide a different level of resolution in remote sensing. Under the direction of John Weymouth, proton magnetometry repeatedly disclosed the presence of earthlodge floors, midden accumulations, storage pits, and fire hearths. Proton magnetometry operates on a simple principle: House depressions and trash deposits at Knife River contain microscopic iron particles. When intensely heated, the particles in burnt houses and fire hearths orient toward magnetic north like a million tiny compass needles. Such anomalies are also created when earth used to refill storage pits has a different composition than the surrounding matrix soil (and thus slightly higher or lower magnetic susceptibility). If this difference is large enough to be measurable, magnetometry can pick up the anomaly.

Weymouth relied on devices called proton precession magnetometers to measure the strength of magnetism between the earth's magnetic core and a sensor controlled by the archaeologist. By analyzing hundreds of these readings taken across a systematic grid, a computer plotter can generate a magnetic contour map reflecting both the shape and the intensity of magnetic anomalies beneath the ground surface.

When everything works just right, the magnetometer can tell archaeologists what is going on beneath the earth's surface. Many archaeological features have characteristic magnetic signatures—telltale clues that hint at the size, shape, depth, and composition of the archaeological objects hidden far below. Shallow storage pits and graves, for instance, have a magnetic profile vastly different from, say, a buried fire pit or a house floor.

The magnetic survey program at Knife River has enabled archaeologists to pinpoint many additional house foundations, augmenting the information gathered by viewing the surface. The use of noninvasive, nondestructive technology has created an innovative data base in which these "magnetic houses" can be inventoried and studied while remaining buried and preserved for the future.

This approach underscores a recent revolution in American archaeology. A century ago, archaeologists simply blasted away at their sites, leaving ruined ruins in their wake. To archaeologists at midcentury, the greatest technological revolution was the advent of the backhoe as a tool of excavation. Today we view archaeological sites differently. Part of this new conservation ethic reflects the definition of archaeological remains as nonrenewable resources. Equally important has been the development of noninvasive technology for relatively nondestructive archaeology. Using the archaeological equivalents of the CAT scan, archaeologists can now map subsurface features in detail without ever excavating them. And when it does become necessary to recover samples, we can, like the orthopedic surgeon, execute pinpoint excavations, minimizing damage to the rest of the site.

Visitors would enter such an earthlodge through a covered wooden entryway, passing stacks of firewood and, after Euro-American contact, maybe even a small corral where a few favorite ponies might be kept safe from marauding enemies. Straight ahead was the central fire hearth, a circular basin 3 to 4 feet (1 to 1.5 meters) across, dug into the floor. Women's cooking and tanning equipment dangled from the center posts. To the left were low storage platforms and tables for preparing food, raised beds, and secure spaces for

George Catlin's overview of the lower Mandan village at Fort Clark, on the upper Missouri River (1837–1839). In each lodge lived several matrilineally related families. Note the skin-covered bull-boats, stored on the domed earth roofs. The poles display scalp trophies and lengths of trade cloth.

personal possessions. The right side was similar, with perhaps the addition of a small sweatlodge. Toward the rear was a sacred space, a shrine for storing the medicine bundle.

Around the perimeter, both inside and outside, were buried food caches, often expanding at the bottom into a characteristic bell shape. These spacious underground chambers, lined with willow branches and grass, held dried meat and garden produce (mostly maize, squash, and beans), sometimes stored in individual leather bags. Eventually, after these cache spaces turned moldy, they were filled with household trash. Here and there were small "pocket caches," buried tool kits containing bone digging tools, hide-working implements, and flakers for flintknapping stone tools into final form.

Trash middens tended to accumulate outside the earthlodges, piling up as the lodges were built and rebuilt. A lodge with well-maintained earthen cover and thatching might last a decade before the structural beams began to rot. Finally, though, it had to be dismantled. This was done with great ceremony, the logs and other wooden members being distributed to helpers. A new lodge would often be built on the spot, again with appropriate formality and ritual.

Plains Village Lifestyles

Knife River was home primarily to the Hidatsa people, who lived there at least as early as 1300, perhaps 1100. They remained in residence until 1845, when disease and warfare decimated them. The Hidatsa shared a basic lifestyle, economy, and religion with the Mandan people, whose traditional territory was downstream from the Knife River area. Even though these two tribes derived from different linguistic stocks and spoke different languages they would eventually join together in a spirit of mutual support.

Summer villages, ordered communities containing as many as 120 earth-lodges, were built on the natural terraces above the Missouri River and its tributaries. The villages often were strategically located for defense on a terrace margin with steep bluffs on two sides and a protective palisade on the third. In wintertime the inhabitants moved into smaller lodges located along the bottomlands, where they were protected from the icy winds and had plenty of firewood at hand.

Although these Knife River villagers hunted bison whenever possible, they relied on a basic farming lifestyle, raising squash, pumpkin, beans, sunflowers, and a tough, quick-maturing variety of corn in their rich floodplain gardens. They celebrated the first corn of summer with the Green Corn ceremony. Berries, roots, and fish supplemented their diet.

The Knife River people were never totally isolated. The upper Missouri River was a lifeline winding across the Northern Plains. Situated as they were, the Knife River people often served as intermediaries between other groups, and they dealt in a wide variety of goods: Knife River flint from quarries west

Sketch by Gilbert Wilson showing the spatial structure of a twelve-post Mandan earthlodge at Knife River, as described by Buffalo-Bird-Woman. This particular lodge was that of Small-Ankle, her father, as it appeared about the time of her own marriage.

of the villages, obsidian from Wyoming, Great Lakes copper, shells from the Gulf of Mexico and the Pacific Northwest. For centuries they maintained these long-distance trade relationships and negotiated tribal alliances with their neighbors.

With the advent of Euro-American exploration, a brisk trade network arose. The outsiders brought colored beads, metal implements, cloth, tobacco, and guns that they gladly exchanged for furs, which became a new form of currency on the upper Plains. As a result of such European contact, the Knife River people began trading in firearms and horses.

The Knife River villages attracted growing numbers of non-Indian traders who inadvertently brought smallpox and other deadly diseases with them. The first historically recorded smallpox epidemic struck the Knife River villagers in 1780–1781, but others had undoubtedly preceded it. Before the 1781 epidemic, the Hidatsa population probably ranged between 4,000 and 5,500 people. In 1798, when explorer and fur trapper David Thompson visited, only about 1,700 Hidatsa survived at Knife River. Another devastating epidemic struck in 1837–1838, killing more than two-thirds of the Hidatsa people. The Mandan suffered even greater losses.

Each epidemic was followed by considerable shuffling of village locations as people grouped and regrouped, combining forces for refuge and mutual defense. After the 1838 epidemic, the weakened Mandan and Hidatsa villages became easy prey for Sioux raiders. Warfare became endemic, and the villagers were unable to tend their gardens or venture out as hunting parties. Little by little, they abandoned the traditional villages. By 1845 the Mandan and Hidatsa survivors from Knife River had joined forces permanently, establishing Like-A-Fishhook village some 40 miles (65 kilometers) upstream and leaving Knife River forever. The remaining Arikaras joined them in 1862; Like-A-Fishhook village was their last attempt to maintain traditional lifeways. In the 1880s, the villagers abandoned the Like-A-Fishhook settlement and moved onto individual allotments at the Fort Berthold reservation. In 1934, the Hidatsa, Mandan, and Arikara were formally united as the Three Affiliated Tribes, and they survive as such today.

Origins: Traditional Hidatsa Views

The Hidatsa people consist of three distinct ethnic subgroups: the Awaxawi Hidatsa (pronounced Ah-WAH-ha-WEE, meaning "Village on the Hill"); the Awatixa Hidatsa (pronounced Ah-WAH-ti-HA, meaning "Village of the Scattered Lodges"); and the Hidatsa proper ("Hidatsa" means "People of the Willows"). Oral tradition suggests that each subgroup had its own origin.

The Awaxawi Hidatsa believe that they moved in from a faraway underground world near Devil's Lake in North Dakota. Their tradition says that they were the second Hidatsa group to settle at Knife River, arriving after the Awatixa. Tribal tradition holds that the Awaxawi people moved westward because of population pressure and warfare with the Cree and Assiniboine people.

The Hidatsa proper maintain an early traditional history closely related to that of the Awaxawis, including an emergence near Devil's Lake and migration northward. They believe that they arrived in the Knife River soon after the Awaxawi. When historical documentation began, the Hidatsa proper were living at the large, fortified Big Hidatsa village.

Awatixa Hidatsa tradition holds that they have always lived near where the Knife and Missouri rivers join, having arrived there well before the other two Hidatsa subgroups. Their origin tale, handed down from generation to generation, begins with a supernatural being who lived in a village in the clouds. Although everyone in his tribe also enjoyed an extraordinary range of metaphysical powers, hunters complained of insufficient game to feed their families. One day, our hero heard the braying of many buffalo from below. Peering through a hole in the clouds, he could see the earth, where boundless herds of buffalo roamed the Great American Plains. Wanting to settle there, he transformed himself into an arrow and soared downward from the sky. He stuck fast and was attacked by an evil adversary named Fire-Around-the-Ankle. Set afire by this attack, our hero was badly burnt as he struggled to free himself; from then on, he was known as Charred Body.

After he freed himself, Charred Body settled on a rise overlooking the Missouri River, where he constructed thirteen earthlodges. Returning to the clouds, he recruited thirteen young couples to settle the new land. Bringing with them seed corn and the necessities for life below, the pioneers—transformed into downward-shooting arrows—accompanied Charred Body to the new village. Awaxawi Hidatsa tradition holds that epic struggles took place between the supernaturals and the villagers who already lived in this new land (perhaps early Mandan people). Charred Body's village prospered. The pioneering families intermarried and the population soared, eventually becoming the thirteen clans of the Hidatsa people.

Awatixa Hidatsa legend states that this first village of thirteen lodges was established on a river terrace overlooking Turtle Creek, about 2 miles (3 kilometers) below the present town of Washburn. Because of the legend, this creek is also known as Burnt or Charred Body Creek.

Digging for Tribal Traditions

Some American archaeologists dismiss Native oral tradition out of hand, believing that the disruptions and population decimation of the historic era created a dramatic break separating the "prehistoric" and "modern" Indian worlds. One archaeologist has expressed this viewpoint in a particularly vivid oxymoron: "I don't think that . . . oral traditions are worth the paper they're written on."

Archaeologists working at Knife River have always taken a contrary view,

Paired views of Sakakawea Village, Knife River: (right) George Catlin Painting and (below, right) aerial view of the village

recognizing the close intertwining of Mandan and Hidatsa oral tradition on the one hand and archaeological exploration on the other. The very first archaeologists working here in the 1880s concentrated on mapping the major village sites. When Orin G. Libby of the newly formed State Historical Society of North Dakota initiated more systematic efforts in the early 1900s, he knew that the Mandan and Hidatsa had abandoned Knife River only a few decades before and that their descendants could provide valuable insights about the archaeological remains. Libby commissioned Sitting Rabbit, a Mandan, to construct maps of all known villages, drawing on information from tribal elders and other Native people.

Over the next century, archaeologists emphasized the degree to which ethnography, ethnohistory, and archaeology provide complementary data sets for understanding tribal origins. Some intriguing evidence has recently emerged indicating a previously unsuspected degree of convergence between archaeological and traditional records of the Hidatsa Indian people. To illustrate how this approach works, let us briefly return to the Awatixa Hidatsa origin story summarized above.

During their archaeological survey of the 1940s, archaeologists George Will and Thad Hecker located an archaeological site near the spot where Charred Body supposedly established the initial Awatixa Hidatsa settlement. Aware of the legend, they called it "the Flaming Arrow site." More than a century ago, W. Matthews had reported seeing some "vestiges" of large circular lodges in the same area. Fifty years ago, ethnographer Alfred Bowers noted that thirteen lodge outlines were visible at the Flaming Arrow site and that Awatixa Hidatsa traditions said these represented the original dwellings of the thirteen distinct lineages of the ancient clan system before the smallpox epidemics. But little concrete archaeology was conducted at Flaming Arrow, and the site was damaged badly by highway and railway construction.

When University of North Dakota archaeologists tested the Flaming Arrow site in 1983, they found some very unusual things buried there. First of all, the house they excavated was oval, quite different from the rectangular or circular houses known from other village sites in the area. It was also curious that corn storage pits—ubiquitous in other such sites—were entirely absent at Flaming Arrow. The pottery was also quite different, cord-roughened and similar to more ancient Woodland ceramics. Radiocarbon dates processed on charred timbers from this house dated about A.D. 1100, making Flaming Arrow the oldest Plains Village site known for this region.

Although the excavations have been quite limited, archaeologists familiar with the evidence emphasize the unique character of the Flaming Arrow site. Both the archaeological record and Awatixa tradition tell basically the same story about the earliest Hidatsa settlement of the Missouri River. Clearly, by 1200 the Plains Village lifeway became firmly established in the Knife River area; today, archaeologists recognize the importance of traditional knowledge by employing the term "Charred Body complex" to describe this early presence, which apparently reflects the presence of both ancestral Mandan and Awatixa Hidatsa people.

Archaeologists Stanley Ahler, Thomas Thiessen, and Michael Trimble summarize the situation at Knife River this way in *People of the Willows*: "Data from several sources indicate that the Hidatsa tribe and their ancestors were most responsible for the prehistoric record . . . at Knife River. The origin traditions of the Hidatsa people tell us something of these early events. . . . This oral history has recently been confirmed and embellished by scientific investigations" (27).

Further Reading

A extraordinary introduction to the archaeology of the Knife River area is *People of the Willows: The Prehistory and Early History of the Hidatsa Indians* by Stanley A. Ahler, Thomas D. Thiessen, and Michael K. Trimble (Grand Forks: University of North Dakota, 1991); see also *M-E ECCI AASHI AWADI: The Knife River Indian Villages* by Noelle Sullivan and Nicholas Peterson Vrooman (Medora, ND: Theodore Roosevelt Nature and History Assocation, 1995).

More general presentations include Douglas J. Lehmer's *Introduction to Middle Missouri Archeology* (National Park Service, 1971); *Like-A-Fishhook Village and Fort Berthold, North Dakota: Introduction to Middle Missouri Archeology* by G. Hubert Smith (National Park Service, 1972); and *The Phase I Archeological Research Program for the Knife River Indian Village National Historic Site, Parts I–IV* by Thomas D. Thiessen (National Park Service, Midwest Archeological Center, 1993).

Further Visiting

At the Knife River Indian Villages National Historic Site (north of Stanton, North Dakota), visitors can see the remains of three Hidatsa Indian villages, with visible lodge depressions, cache pits, and fortification ditches.

EIGHTEEN

Little Bighorn Battlefield

25 JUNE 1876

Sioux, Cheyenne, Crow, and Euro-American

cultures meet at Hardin, Montana

The United States of America was still celebrating its hundredth birthday when the shattering news arrived. The 6 July 1876 edition of the *New York Herald* headlined the story:

A BLOODY BATTLE.
An Attack on Sitting Bull on the Little Horn River.
GENERAL CUSTER KILLED.
An Entire Detachment Under His Command Slaughtered.
SEVENTEEN OFFICERS SLAIN.
Narrow Escape of Colonel Reno's Command.
A HORRIBLE SLAUGHTER PEN.
Over Three Hundred of the Troops Killed.

Incredibly, a combined force of Sioux (Lakota) and Cheyenne Indians had annihilated half of Lieutenant George Armstrong Custer's invincible Seventh U.S. Cavalry. Custer had fallen victim to one of the largest fighting forces ever seen on the Plains. Never mind that the U.S. government had violated its treaties with Native Americans—there was gold in them thar (Black) hills. The railroad must be built, and savages could not be allowed to impede the path of progress.

The gunsmoke had barely lifted at the Little Bighorn before moral outrage was tempered by morbid curiosity. Hundreds of volumes have been written about the Battle of the Little Bighorn, and hundreds of thousands of people flock each summer to that windy, grassy field, where National Park Service guides struggle with still-unanswered questions.

What was Custer trying to do? Some think that he was attempting to surprise the Sioux enemy; others claim that he was scrambling to resurrect a foundering military career by stealing the glory from his commanders. Some skeptics view his headlong push to the Little Bighorn as a campaign ploy to gain the presidential nomination from the Democratic National Convention that would assemble in St. Louis on June 27—two days after his death.

Nationwide reaction was no different in 1876 than it is today when bad news arrives: bewilderment, followed by incredulity, rage, and a craving for instant revenge. America became galvanized. Quick-tempered volunteers from all corners of the land offered their services. Salt Lake City wanted to send 1,200 troops. Austin's *Daily State Gazette* insisted that "Texas deserves the honor of attempting to wipe out the Sioux." Not to be outdone, the *Daily Herald* in Dallas countered, "Killing a mess of Indians is the only recreation our frontier rangers want." But an editorial in the taciturn *London Times* struck a different tone, suggesting that Custer's defeat was perhaps deserved.

206
—

Custer's Last Stand (1899) by Edgar S. Paxson. This fanciful nineteenth-century depiction of the Little Bighorn Battle encapsulates the anti-Indian perspective that has traditionally surrounded this historic battle.

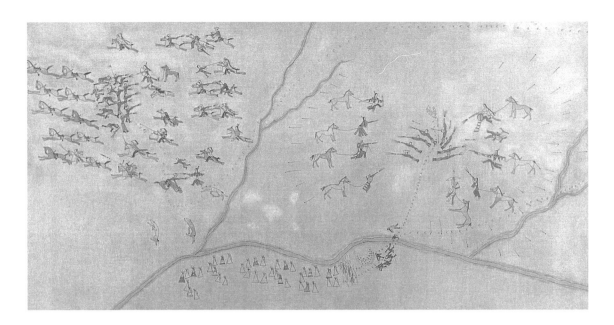

After all, "The conduct of the American Government towards the Indians of the Plains has been neither very kindly nor very wise."

More has been said about the Battle of the Little Bighorn than any other battle in American history, except perhaps Gettysburg. To understand the battle behind the battle, we must look beyond the well-chronicled saga of Custer and his Seventh Cavalry to the broader context.

The Little Bighorn fight took place on 25 June 1876. This is an Indian drawing of the battle.

Some Sioux and Cheyenne Perspectives

The Indian coalition that faced Custer consisted mostly of Cheyenne and Hunkapapa Sioux, the latter led by Sitting Bull. Each group saw the events of 25 June 1876, in a different light, often reflecting their tribal loyalties.

Warfare was nothing new to the Hunkapapa Sioux, who had battled their enemies alone and in concert with allied tribes such as the Northern Cheyenne and the Northern Arapaho. They fought to preserve their hunting grounds. They fought to defend themselves against the incursions of other Indian tribes. They fought for booty, particularly the horses by which they computed their wealth. They fought for retribution, for respect, and to obtain the war honors that defined leadership.

The traditional enemies of the Hunkapapa Sioux were legion. The Sioux had migrated onto the upper Plains from the headwaters of the Mississippi River in the sixteenth century, warring with the Cree, who themselves were being pressured from the east. In Chapter 17 we saw that by the mid-nineteenth century the sedentary farming tribes along the Missouri—the Arikara, Mandan, and Hidatsa—had been so devastated by smallpox that they could no longer impede the advances of their traditional Sioux enemies. But as the

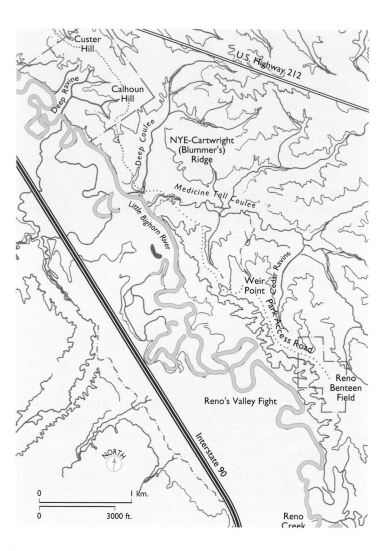

Hunkapapa moved westward into Crow territory, they were vigorously resisted by those who had been there first. Sioux and Crow hunting territories now overlapped, creating a deadly zone of conflict on the Powder River. Sioux war parties also ventured northward, engaging the Assiniboine, each raiding far into the other's territory. Such intertribal warfare and skirmishing eventually allowed a young Sitting Bull to solidify his position of leadership. Throughout much of his life, he wore a single eagle feather, symbolic of bravery in battle—not against the hated whites but against the long-standing enemy, the Crow.

When the Fort Laramie Treaty of 1851 tried to establish peace between the United States and all the Plains tribes, it also required that the tribes make peace with one another. As various tribal representatives signed, smug American onlookers naively assumed that the signatories understood and agreed to these conditions. In truth, the Sioux delegates could agree on few issues even among themselves, and sometimes they violently split into factions. Few Americans understood the depth of these long-standing hostilities and the degree to which intertribal warfare was ingrained.

The Hunkapapa had only recently wrenched the land around the Yellowstone River and its tributaries from the Crow, at the cost of considerable Sioux blood. This land now defined the Sioux homeland, and they had no intention of stopping the fight with their Indian enemies. Bloody attacks and counterattacks continued, often ending only after hand-to-hand fighting had dispatched the last warrior. Even on that warm June evening in 1876, as they camped in the valley of the Greasy Grass, Hunkapapa leaders worried less about a confrontation with American soldiers than about their traditional Indian enemies.

Some Crow Perspectives

The Sioux's greatest enemy, the Crow, were descended from the settled Plains farming villages. At some point, farming Hidatsa people from the Knife River area began to overwinter on the Plains, tracking elk, antelope, and buffalo. The Crow trace their ancestry to one of these farming groups that eventually gave up settled village life to hunt full time on the Plains.

Before long, outsiders began encroaching on this new Crow territory. The Oregon Trail opened in 1847, and the gold rush of 1849 unleashed a flood of Euro-Americans westward. Sioux and Cheyenne were being forced out of their own territories by non-Indian settlers, and whites in northwestern Montana forced the Blackfoot southward until they encroached upon the western reaches of Crow territory. The nineteenth-century Crow people became understandably protective of their hunting territories. According to historian Fred Hoxie in *The Crow* (1989), "A century earlier they might have allowed a visit from a Sioux or Blackfoot band, [but] in 1850, the appearance of strangers meant war (59)."

Faced with the prospect of increased warfare and more Euro-American migration across the Plains, the Crow readily joined leaders of other Plains tribes at Fort Laramie in the summer of 1851 to clarify their boundaries. Thousands of Crow, Sioux, Blackfoot, Shoshone, Cheyenne, and others—the largest gathering of Plains Indians ever recorded—sat down to iron out their differences. When it was over, the Crow homeland had been precisely, and artificially, delineated on official government maps. It was bounded on the east by the Powder River, on the west by the headwaters of the Yellowstone River, on the north by the Missouri and the Mussellshell rivers, and on the south by the Wind River Mountains. Although it was huge, covering 38 million acres (15 million hectares), this new Crow country excluded many places the Crow traditionally considered their own.

Warfare had always been an integral part of Crow life. Now they chose to side with the United States, following this difficult course because they perceived that it would best serve their own tribal interests. The Crow well understood that most of the Indian tribes menacing them had themselves been pushed out of their homes by white settlers elsewhere in the United States; they looked to their U.S. allies to live up to the treaties and deliver them from the Indian newcomers.

Although suspicious of white incursions, the Crow never went to war with the United States. Many Crow deliberately chose neutrality, withdrawing into the Bighorn Mountains to avoid hostilities. Other tribal members elected to serve actively with their U.S. allies, mostly as army scouts and couriers. As the hostilities at the Little Bighorn remind us, the "Indian Wars" in the 1860s and 1870s were not simply battles between whites and Indians. The Crow were trying to protect their territory from all manner of land-hungry enemies— not just white miners, settlers, and soldiers, but Indian enemies as well.

Crow scouts rode with Custer's ill-fated Seventh Cavalry into the valley of the Little Bighorn. But following the custom of the time, Custer discharged them before the fighting began. Most left the battlefield, herding captured horses, heading home. But several Crow warriors did remain, two being wounded in action with Major Marcus Reno's detachment. After the battle U.S. forces pursued the Sioux and eventually forced them to return to Dakota Territory. The Crow quickly reoccupied that portion of their lands upon which the Sioux had been forcibly encroaching for two decades.

Harassed on all sides by powerful enemies, Crow leadership maintained a steady course. Although some sided with the United States and others pledged neutrality, their unswerving goal was always to maintain their homeland. Pursuing their own brand of statesmanship, the Crow suffered far less than most Plains tribes during this violent era in American history.

An Anglo-American Bias

Native American eyewitness accounts have, of course, been available since the battle took place, but American history has traditionally dismissed such narratives. After all, when Indian contestants were first interviewed, many battlefield recollections were tempered by a fear of retribution for Custer's defeat. Given the mood of the country at the time, this fear was clearly warranted. And because of the freestyle conduct of Indian warfare—"unencumbered by American Victorian standards of military obedience and duty"—Native American warriors saw little of the overall picture of the battle.

Particularly disturbing to Custer partisans were the Indian accounts that attributed cowardly behavior to soldiers of the Seventh Cavalry: breakdowns in military discipline, widespread panic, attempted desertions, and even suicide. Many Indian eyewitnesses spoke in disgust of the troops' hysteria, suggesting that unbecoming cowardice had tarnished the image of a brave adversary. Even today, American historians sometimes discredit Indian testimony as suspect, garbled, and overly partisan.

210
—

Wildfire Archaeology at the Little Bighorn

Today another perspective can be brought to bear. In 1983 an accidental grassfire spread across the national monument, clearing the terrain of brush and grass and making possible an unprecedented view of the battlefield. During his initial walkover after the fire, archaeologist Richard Fox turned up a variety of artifacts, including human remains, cartridge cases, and buttons. Additional archaeological research was clearly in order.

Historians of the Little Bighorn have long recognized the importance of plotting artifact locations as a key to understanding the progress of the battle. But before the grassfire, such locational evidence was spotty and unreliable, coming mostly from indirect accounts by relic collectors. It is not very useful to know that a particular artifact was found on Last Stand Hill; historical archaeologists needed to know the precise find location—including depth and orientation in the ground—to establish its contextual relationships with other artifacts. The fire changed all that.

Knowing that most military hardware was metallic, the archaeological team began with metal detectors. They superimposed a permanent grid system over the entire 760-acre (310-hectare) battlefield. Skilled operators then worked systematically across the area. Avocational archaeologists were invaluable here, supplying their own metal-detectors and volunteering their time. The metal-detector operators worked at 15-foot (5-meter) intervals, each sweeping an arc of 5 to 7 feet (1.5 to 2 meters). When an artifact was found, the location was pin-flagged and a field-specimen number assigned. A follow-up crew collected surface specimens and carefully hand-excavated buried artifacts. The sweep turned up more than five thousand artifacts. Archaeologists also conducted excavations around the bases of the numerous marble markers that today dot the battlefield landscape, searching for the clothing, equipment, and human bones that would confirm a soldier's final position.

We now have a much clearer picture of what occurred on 25 June 1876, what kinds of weapons the Indians carried, exactly where many of the troops fought, how they died, and what happened to their bodies after death. We also know precisely how the troops were deployed, what they wore, and how they fought. Here are some highlights.

The Mystery of the Deep Ravine

Both white and Indian accounts indicate that twenty-eight soldiers (presumably from Company E, Seventh Calvary) were cut off, killed, and buried in a rugged gully that runs across the sloping plain below Last Stand Hill. The circumstances of this skirmish are unclear. Were these men part of an orderly defense? Or did they sense defeat and try to hide? The accounts vary, but all agree that the missing men were killed in the rugged coulee today known as the Deep Ravine.

Three days after the battle, surviving members of Major Reno's command and some members of the relief column identified whatever remains they could, covering the bodies with brush and mounded dirt. In 1877 and 1879, military personnel were assigned to gather exposed remains and rebury them in a mass grave on top of Last Stand Hill where a granite memorial shaft stands today. But the bodies in the Deep Ravine were so decomposed that a hasty mass grave was dug near where the men fell. It is unclear whether the burial detail of 1877 or the reburial party successfully located the grave, and the precise burial spot has been lost to history.

The metal-detector survey turned up little in the area of the Deep Ravine, and subsurface testing was equally unproductive. Because historical accounts universally identified the Deep Ravine as the final destination of the missing troopers, the excavators concluded that the human bones and artifacts are probably too deeply buried for detection by conventional archaeological procedures.

The archaeological team then brought in geomorphologist C. Vance Haynes, who developed a stratigraphic sequence of the various sediments in the Deep Ravine. Haynes found that Deep Ravine had been substantially deeper than it is now; historic-period fill and debris was at least 14 feet (4 meters) deep in places. Stratigraphic evidence showed that, provided the remains have not already washed away, they lie too deep for metal-detection technology.

Haynes suggests that the burial detail, in the process of excavating and filling the mass grave, may have created a low check-dam across the ravine. If the bodies were tightly clustered and heaped into piles, they could have caused sediments to build up and partially fill up the ravine floor. If this scenario is correct, then the bodies of the missing troopers lie deeply buried below the water table. This would mean that—if the water table has remained stable over the past century—bones, cloth, metal, and other materials buried there may be quite well-preserved. At present, Haynes's proposed solution to the mystery of the Deep Ravine remains an untested, if intriguing, hypothesis.

Excavating the Marble Markers

In 1890, distinctive marble markers were set up to commemorate locations where individual soldiers fell. Two hundred and fifty-two marble markers stand on the battlefield today, indicating where roughly 210 men from Custer's command died. Contemporary researchers do not know why there are 42 more markers than there were soldiers killed in the fight.

In the 1984–1985 investigations, archaeologists tested about 15 percent of the marble markers, employing random sampling methods where feasible. Excavation units 6 feet (2 meters) square were laid around the markers targeted for testing. All fill was carefully hand-excavated and screened. Human remains and battle-related artifacts were found in nearly all of the marker excavations.

These human bones, studied by forensic anthropologist Clyde Collins
Snow, provide new insights about the battle and those who died there. The
recovered remains are mostly isolated teeth and smaller bones—body parts
easily overlooked by the reburial parties who transferred skeletons from their
original graves to the mass grave in 1881.

Several students of the battle have suggested that the battle ended so
quickly because a sizable number of Custer's troops committed suicide once
they realized the battle was lost. But this mass-suicide theory receives no
support from the archaeological evidence: None of the cranial fragments
showed evidence of gunshot trauma, the kind of damage expected in battlefield suicide. Instead, the osteological evidence suggests a relatively brief
firefight, followed by close-range dispatch of the wounded. Most troopers
remained alive, if helpless, and were finished by massive blows to the head.
Consistent with Indian accounts of the aftermath, many troopers were muti-
lated and dismembered after death.

Snow's forensic examination also led him to question the identifications
made by the original burial crews, including those of eleven officers and two
civilians whose remains were shipped to their families in 1877. Similar misiden-
tification also seems likely for the remains of General Custer himself. The
description of Custer's grave and its contents is seriously at odds with the
accounts of the original burial, raising the possibility that the wrong remains
reside today in "Custer's Grave" at the U.S. Military Academy at West Point,
New York.

Microchronology of the Battle

Historians have long debated the exact deployment of troops—where the combatants took up positions and how they moved as the battle unfolded. By recording the precise locations of cartridge cases and thus charting the movements of weapons as the battle developed, researchers could map out the flow of battle.

Artifact distributions make it clear that the final Seventh Cavalry position was a V-shaped formation, along the east side of Custer Ridge and along the South Skirmish Line. This is where most army-related artifacts such as cartridge cases, buttons, spurs, equipment, and human bone are found. These troop positions are further indicated by the presence of impacted bullets from Indian firearms. At least seven discrete Indian positions can be defined on the basis of cartridge-case distributions from weapons known to have been used by the Indians. The archaeological evidence shows that the cavalry was relatively stationary but the Indians moved freely about, overrunning one position after another.

FORENSIC ARCHAEOLOGY: PIECING TOGETHER MITCH AND MIKE

In his initial reconnaisance following the brush fire, Richard Fox found human bones on the surface between Markers 33 and 34. The next year an excavation unit revealed some river cobbles, two bullet fragments, a partial boot, some buttons, and fragments of a human skull, finger, and tailbone. Investigators also found a deteriorated piece of a cedar stake, probably part of a tipi pole placed there by one of the reburial parties to mark the grave site.

Archaeologists and forensic specialists were able to wring an extraordinary amount of information from these meager finds. One of the buttons was mother-of-pearl, suggesting civilian Euro-American clothing rather than military issue. One of the bullets was from a weapon used only by Indians in the battle, suggesting that the victim was associated with the Seventh Cavalry. The bones, all from a single individual, suggest an age at death of 35 to 45 years. The skull fragment displays several traits characteristic of Mongoloid populations, including relatively broad nasal aperture, low nasal sills, and large cheek bones. But the incisors are not shovel-shaped (almost a Mongoloid racial hallmark). The skull can be interpeted as reflecting a Caucasian-Mongolian racial mixture. Taken together, the evidence suggests that Markers 34 and 35 were placed to mark the final location of a middle-aged part-Indian civilian who fought with Custer's battalion.

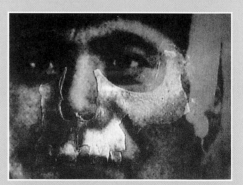

Facial bones from Markers 33–34 (top) and an overlay of bones on a picture of Mitch Boyer (bottom).

The battle ended on Last Stand Hill. Archaeological evidence confirms that firing must have been intense from both sides, with Indian movements continuing along two broad lines converging on the hill. This was the final position defended by remnants of the five companies of the Seventh Calvary that had ridden onto this ridge only a short time before. The combined Cheyenne and Sioux forces overran them.

The archaeological evidence clearly supports Indian accounts of the battle, which are considerably more accurate than the ad hoc reconstructions of the burial details. Archaeologist Richard Fox concludes, "What the archaeology and the Indian accounts told me, is that there was no gallant defense on Custer Hill . . . the troopers put up no substantial resistance. It was a complete rout."

Still Battling at the Little Bighorn

Americans will always love their martyrs, of course, from the doomed Texans defending the Alamo to the star-crossed troopers on Last Stand Hill. Reen-

Battle records show that the only known white and Indian mixed-blood to have died at the Little Bighorn was 38-year-old Mitch Boyer, employed by Custer as a scout and interpreter. The forensic team created a video overlay of the facial bones found at Markers 33–34 with a photograph of Boyer taken shortly before his death. The match is precise. Boyer's father was French; his mother a Santee Sioux. The Sioux considered him a traitor, and Sitting Bull once offered one hundred ponies for his head. On the morning of the battle, Boyer and other scouts were assigned to observe the hostile encampment on the Little Bighorn Valley, some 15 miles (25 kilometers) away. Boyer, a frontiersman of considerable experience, warned Custer that the encampment was the largest he had ever encountered in his thirty years on the Plains. To engage this huge force was to invite certain death. Custer, of course, ignored Boyer's advice, and both perished later in the day.

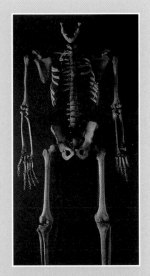

Reconstruction of "Mike"

A nearly complete weathered skeleton, missing only the skull, was recovered in 1985. Although no positive identification could be made— the skeleton was nicknamed Mike— Clyde Snow reconstructed a great deal about his life and painful death. Mike was a white male between 19 and 22 years old. He stood about 5 feet 6 inches (167 centimeters) tall and was right-handed and stocky, with well-developed muscles. He was wounded three times: in the right side of his chest, in the upper abdomen, and just above his left wrist. He had also been stabbed in the left shoulder (or perhaps shot with a metal-tipped arrow). His skull had been crushed by a massive blunt force, probably a rifle butt or war club. It is unclear if Mike was alive when his legs were hacked off at the hip joint.

Snow summarized the evidence in a 1986 *National Geographic* article: "My guess is that maybe 10 percent of the men were killed instantly. The rest lay there wounded but alive. Like Mike. . . . Here's this young kid, this trooper with a sore back and now lying there with those terrible wounds. What was going through his mind? He could hear the Indians yelling. He knew they were coming. That strikes me as particularly horrible (800)."

The final phase of the
Little Bighorn battle and
the destruction of Custer's
command

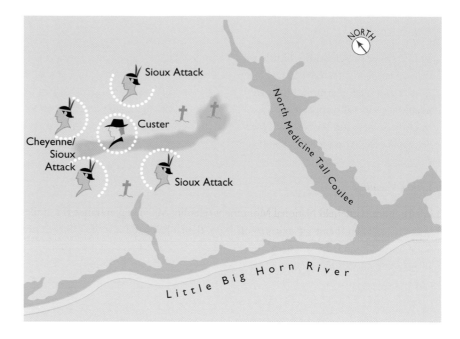

actments of such watershed historical events have become an important way
for Americans to keep their own history alive. Each year since the 1970s the
Hardin chamber of commerce has sponsored a recreation of the Battle of the
Little Bighorn. Striving for authenticity and historical accuracy, the participants portray the events that culminated in Custer's colossal defeat.

Crow people are understandably ambivalent about the Little Bighorn.
The battle took place on their land, and while some of their ancestors scouted
for Custer's cavalry, many Crow wonder if they should have thrown in with
Sitting Bull instead.

Similar feelings surfaced when Barbara Booher, a Native American who
became Park Superintendent at the Little Bighorn, first visited the battlefield
in the late 1980s. Although entranced by the aura of the place, she also noted
that the dozens of white marble grave markers were mostly for American soldiers. Estimates of the Indian dead range between about 40 and 150 killed that
day, and yet only one was commemorated with a grave marker. "It looked like
a No Parking sign," she commented dryly. Two years later, when Booher
took charge, she vowed, "This will all change."

Some things have changed already. On December 9, 1991, President George
Bush signed into law a bill changing the name of the Custer Battlefield
National Monument to the Little Bighorn Battlefield National Monument.
A monument to honor the Cheyenne, Sioux, and other Indians who perished
defending their families, their lifestyle, their culture, and their lands will be
erected near the existing monument honoring the cavalry.

Further Reading

The best single presentation of Little Bighorn archaeology is *Archaeological Perspectives on the Battle of the Little Bighorn* by Douglas D. Scott, Richard A. Fox, Jr., Melissa A. Connor, and Dick Harmon (Norman: University of Oklahoma Press, 1989). Highly readable accounts also appear in Andrew Ward's "The Little Bighorn," *American Heritage* (April 1992): and Robert Paul Jordan's "Ghosts on the Little Bighorn," *National Geographic* 170, no. 6 (1986): 786–813.

Further Viewing

Little Bighorn Battlefield National Monument (Hardin, MT; main entrance is 15 miles [25 kilometers] southeast via exit 510 off I–90, then a half mile east via US 212) not only commemorates the dramatic Indian victory of June 1876 but also embraces a national cemetery established in 1879. Just inside the entrance is a visitor center and museum.

Picture Credits

Neg. no. 6684, American Heritage Center, University of Wyoming, 64 (top); American Museum of Natural History, 21, 23, 25, 31, 33, 34, 35, 36, 40, 68, 87, 89, 97 (bottom), 98, 99, 100, 106, 117, 128, 132, 138, 143 (bottom), 146, 147, 158 (bottom), 166, 170, 177; Transparency no. 5640 courtesy Department of Library Services, American Museum of Natural History, 67; Transparency no. 329206 courtesy Department of Library Services, American Museum of Natural History, 96; Transparency no. 316405 Courtesy Department of Library Services, American Museum of Natural History, 115; Transparency no. 6164 Courtesy Department of Library Services, American Museum of Natural History, 119; Transparency nos. 3512, 5603 Courtesy Department of Library Services, American Museum of Natural History, 134 (top), 135; Transparency no. K15910 courtesy Department of Library Services, American Museum of Natural History, 137; Transparency no. 291229 courtesy Department of Library Services, American Museum of Natural History, 143 (bottom); Bernisches Historiches Museum, 194; Dr. John Blitz, 48; Reproduced with the permission of the Trustees of Boston University and The Journal of Field Archaeology, 19; Buffalo Bill Historical Center, 206; Descriptive Archaeology, vol. 1, Crow Canyon Archaeological Center, 121; Photograph: Dirk Bakker © The Detroit Institute of the Arts, 88; Dr. John A. Eddy, 69; Dr. Jon L. Gibson, 97 (top); Dr. J.L. Giddings, 51; Gilcrease Museum, 164; The Stampede, Fredrick A. Verner, Glenbow Collection, Calgary, Alberta, Canada, 54 (top); Harvard University and Robert Neuman, 173 (top); Holmes, William Henry. Washington, DC: Smithsonian, 1883, 111; Joslyn Art Museum, Omaha, Nebraska; Gift of Enron Art Foundation, 199 (bottom); Kansas City Star, 164; Ruth Kirk, 73, 77, 82; London Museum of Archaeology, 190; London Museum of Archaeology, artist Ivan Kocsis, 185, 186; Makah Cultural and Research Center, 81; Mesa Verde National Park, 122, 124, 125; Missouri Historical Society, 158 (top); L.A. Huffman, photographer Montana Historical Society, 59; National Library of Canada / C-13066, 187; Photo by Charles A. Lindbergh, Courtesy Museum of New Mexico, Neg. No. 130232, 129; National Archives of Canada / C-00403, 54 (bottom); National Museum of American Art, Smithsonian Institution, Gift of Mrs. Joseph Harrison, Jr., 195 (bottom), 200 (top); Courtesy of the National Museum of the American Indian, Smithsonian Institution, 169 (bottom); Courtesy of the National Museum of the American Indian, Smithsonian Institution, neg 4428-1090, 168; Courtesy of the National Museum of the American Indian, Smithsonian Institution, neg. 20815, 169 (middle); National Park Service, 84 (bottom), 85; National Park Service, 113; National Park Service, Dr. J.L. Gidding and Dr. D.D. Anderson, 43, 45, 46, 47, 50; National Park Service, Knife River Indian Villages, 195 (top); National Park Service, Chaco Culture National Historical Park, created by Tom Windes, 116; National Park Service, photo by Russ Hanson, 200 (bottom); Robert Neuman, 173; Douglas Scott, National Park Service, 208, 213, 214, 215, 216; General Research Division, The New York Public Library, Astor, Lenox and Tilden Foundations, 84 (top), 176; Nordenskeld, Gustaf E.A. Stockholm, Chicago: P.A. Norstedt, 1893, 116; Rob Stimpson / Northshore Graphics, 183; Peabody Museum, Harvard University, 109; Don Grey, Plains Anthropologist, 63; Jack Brink, Provincial Museum of Alberta, 56, 58, 60; Pueblo Grande Museum and Archaeological Park, City of Phoenix, 142 (top); From Rowland B. Orr, 1917:50, "Indian Fortified Village Site", in Twenty-Ninth Annual Archaeological Report, Toronto, 189; Harvey Rice, 80; Joe Saunders, 103; Smithsonian Institution, photo by Dache Reeves, 107; Soil Systems Publications in Archaeology No.20, Vol. 4, 1994. FS 17161, FS 24427, 149, 154; Squier, E.G. and E.H. Davis. Washington DC:Smithsonian, 108; Andrew M. Tomko III, 55; Courtesy of the Illinois Transportation Archaeological Research Program, University of Illinois, 159; ©The University of Iowa, Midcontinental Journal of Archaeology, 110; Vincas Steponaitis, Research Laboratories of Archaeology, The University of North Carolina at Chapel Hill, 174; Stovall Museum, University of Oklahoma, 169 (top); University of Pennsylvania Museum, 20; Washington State Historical Society, Tacoma, 72; Ozette Archaeological Research Reports, vol. 1. House Structure and Floor Midden, Washington State University, 77 (top), 78; West Point Museum Collection, United States Military Academy, 207

Index

Note: Page numbers in boldface indicate illustrations.

Backwater fishing, 103, 104
Baegert, Johann Jakob, 38
Bainbridge, OH, 86
Baha California, 38
Ballcourts, Hohokam, 145
Barrow Pit Locality, 20
Barter, Hohokam, 143
Basketmaker Caves, AZ, 48
Basketmaker farmers, 87
Basketmaker I stage, 118
Basketmaker II stage, 118,
    119, 120
Basketmaker III stage, 118,
    120, 130
Bat guano, 32, 33
Battle of Little Bighorn, 13
    anti-Indian perspective on,
    206
    and forensic archaeology,
    14
    Indian drawing of, **207**
Bayou Macon, **96**, 99
BBB Motor site, 159
Beach ridges, Cape
    Krusenstern, 42, 44–45
Beans, 120, 182, 199
Bedford Mound, IL, **88**
Bering Strait, 28, 42, 49
Big Hidatsa village, 193, **195**,
    201
Big Horn Basin, 68
Big Horn Medicine Wheel
    as ancient observatory,
    68–70
    as ancient tomb, 65
    scientific exploration of,
    62–65
    summer solstice at, **69**
    as vision quest site, 66–68
Bighorn Mountains, 210
Big Horn Range, WY, 63
Big House, 141, 144
Binford, Lewis, 35
Bioarchaeology, 14
Bird, Junius, 56, **96**
Bird bone whistle, **33**
Birger figurine, **159**, 160
Birnick tradition, 45, 46
Bison, 20, 24, 25
    first discovery of, 2–3
    how used, 59
    hunting, 26, 27, 49, 53,
    199
    jumps, 52
    long-horned, 28
    skull, **67**
    See also Buffalo; Buffalo
    jumping
Blackfoot people, 52, 54, 57,
    209
Black Warrior River, AL,
    172, 173
    valley of, 177, 178

Woodland occupation in,
    175
Blackwater Draw, NM, 7
    current views on, 25–27
    fossil areas of, 20
    as hunting ground, 25
    Locality No. 1, 20
    location of, 19–20
    stratigraphic diagram of, **24**
Blitz, John, 48
Blower, electric, 34
"Blowouts," 20
Bodmer, Karl, 194
Bow and arrow, in America,
    48
Bowers, Alfred, 203
Bowhead whales, 46
Boyer, Mitch, 14, 215
Bradley, Bruce, 139
Brain, Jeffrey, 163
Braun, David, 179
Brink, John W., 57
Brown, James A., 167
Brush Creek, OH, 106
Buffalo, uses of, **59**
    See also Bison
Buffalo berries, 40
Buffalo-Bird-Woman, **199**
Buffalo jumping, art of,
    53–**56**
Buffalo Rift, A (Miller), **54**
Buffalo runners, 55
Building mounds
    significance of, 103
    See also Mounds
Bullboats, 197, **198**
Bulrush(es), 37
Burden baskets, 96–97
Bureau of Indian Affairs, 76
Bureau of Reclamation, 10
Burial forms
    Hohokam, 149
    Mound, 159, 172, 179
Burial mounds, Mexican, 87
    See also Mounds
Bush, George, 216
Byrne, Roger, 183–84

Caddo Indians, 95, 171
Cahokia, IL, 10, 96, 155
    demise of, 161–62
    derivation of term, 152
    kin-based societies in, 154
    stone goddesses of, **159**–60
    woodhenge structures of,
    156
    See also American Bottom
*Cahokia and the Archaeology of
    Power* (Emerson), 159
Cahokia Creek, **154**
Cahokia Mounds State
    Historic Site, 157
Calumet, 88

Camel, 25, 27
Cameron, Terry, **110**–11
Canadian Museum of
    Civilization, 188
Cannibalism, 123, 190
Cape Krusenstern, 7, 42–51,
    aerial view of, **43**
    excavated settlements of,
    43
    gravel deposits at, 44–45
Carbowax, 79
Caribou, hunting of, 46, 49,
    50
Carson Desert, 30, 31
Casa Grande, 150
Cather, Willa, 116
Catlin, George, 194, 196,
    198, 200
Cattail pollen, 37
Cedar plank house, 75, **77**
Cedar pole litters, 168
Center for Desert
    Archaeology, 144
Central Chamber, 166
Ceramic stylistic dating, 144,
    176
Ceremonialism, Hohokam,
    149–50
Chaco Canyon, NM, 91,
    123, **128**
    abandonment of, 139
    Ancestral Pueblo culture
    at, 127–40
    construction in, 139
    evolving architecture of,
    128–29
    kivas of, **136–37**
    phenomenon, 137–38
    population estimates of,
    137
    road system of, **134–36**
Chaco National Monument,
    135, 139
Chaco Wood Project,
    131–34
Chacra Mesa (*Tzak aih*), 127
Charnel house(s), 85, 167
    See also Great Mortuary;
    Mortuary(ies)
Charred Body Creek, 202
Charred Body story, 13, 202
Chetro Ketl, 128, **137**
Cheyenne culture, at Hardin,
    MT, 205–17
Cheyenne Sun Dance Lodge,
    63
Chiefdom(s), in American
    Bottom, 152, 162
"Chief's Room," 165
Child burials, 179
Chillicothe, OH, 83, 86, 92
Choctaw people, 153, 178
Choris culture, **47**, 49

Chukchi Sea, 50
    beach ridges on, 44–45
Chukotka coast, 42
Chuskas Mountains, 133
Cienega phase sites, 144
Cist(s), 120
Civilian Conservation Corps,
    174
Clan affiliation, importance
    of Iroquoian, 187
Claplanhoo, Ed, 75
Classic Maya, 2
Clay balls, Poverty Point, **99**,
    102
Clements, Forrest E., 166
Cliff dwellers, Mesa Verde,
    124
*Cliff Dwellers of Mesa Verde,
    The* (Nordenskiöld), 117
Cliff Place, discovering,
    114–17, **115**
Clovis culture, 3, **23**, 24–**25**
    discovery, 19–23
Clovis Fluted, 24
Clovis point(s), **21**, 22, **23**,
    24, **25**
    See also "Folsom-like"
    points
Coast Alaskan Chronology,
    defining, 45–50
Coast Guard, 77–78
Cognitive archaeology, 62
Collins, Henry, 42, 44
Colorado, Ancestral Pueblo
    culture in, 114–26
Colorado Museum of
    Natural History, 2, 3
Columbian mammoths, 21,
    24, 25
    See also Mammoths
Columbia River, 81
Commoners, at Mound 72,
    159
Connolly, Robert, 87
Coprolites, 33, 37–38
Corn, Knife River, 199
    See also Maize agriculture
    (horticulture)
Corn Mother tales, 136, 160
Corps of Engineers, 167
Corral, at Head-Smashed-In,
    56, 58
Cotter, John, 21–23, **22**, 25
Covenant(s), 66
Craig Mound, 164
    Great Mortuary at, 167–69
Crawford Lake area, **183**
    as meromictic, 183
Creation tales, 12
Cree people, 201, 207
Creek people, 153, 178
Cremation, 145, 149
Crematory, 167

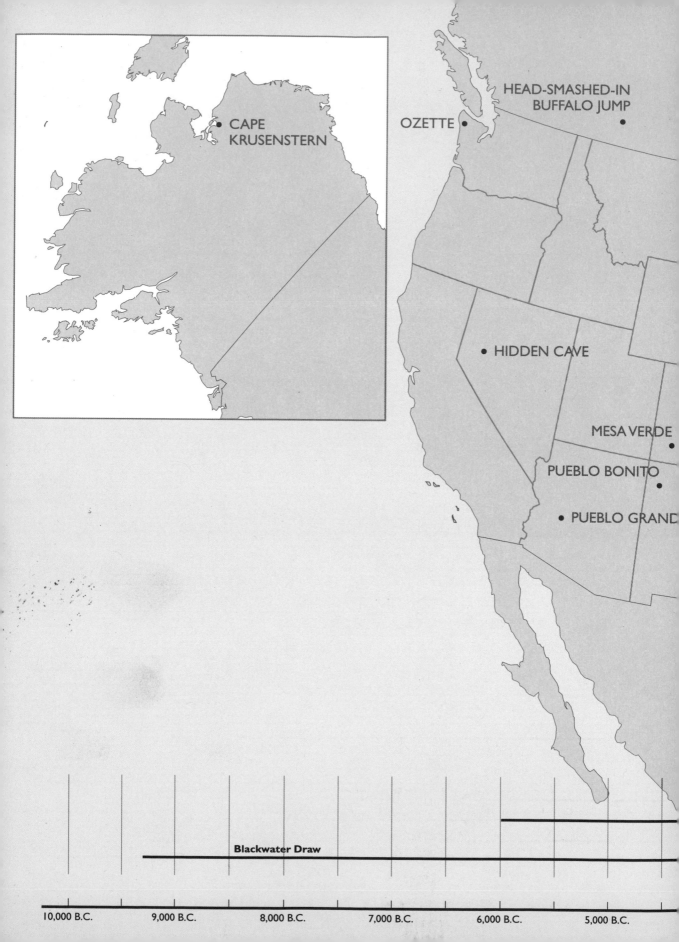

CAPE
KRUSENSTERN

OZETTE •

HEAD-SMASHED-IN
BUFFALO JUMP

• HIDDEN CAVE

MESA VERDE

PUEBLO BONITO

• PUEBLO GRAND

Blackwater Draw

10,000 B.C.   9,000 B.C.   8,000 B.C.   7,000 B.C.   6,000 B.C.   5,000 B.C.